THE UNZIPPED GUIDES™

for everything you forgot to learn in school

Peterson's

Short Stories
UNZIPPED

Ceil Cleveland

PETERSON'S

A **(n)elnet** COMPANY

PETERSON'S

A ⓝelnet COMPANY

About Peterson's, a Nelnet company

Peterson's (www.petersons.com) is a leading provider of education information and advice, with books and online resources focusing on education search, test preparation, and financial aid. Its Web site offers searchable databases and interactive tools for contacting educational institutions, online practice tests and instruction, and planning tools for securing financial aid. Peterson's serves 110 million education consumers annually.

For more information, contact Peterson's, 2000 Lenox Drive, Lawrenceville, NJ 08648; 800-338-3282; or find us on the World Wide Web at: www.petersons.com/about.

Editor: Fern A. Oram; Production Editor: Susan W. Dilts; Manufacturing Manager: Ray Golaszewski; Composition Manager: Gary Rozmierski

ISBN-13: 978-0-7689-2481-7
ISBN-10: 0-7689-2481-2

Printed in the United States of America

10 9 8 7 6 5 4 3 2 1   09 08 07

First Edition

CREDITS

Reprinted from Edgar Allan Poe, *The Tell-Tale Heart*.

Reprinted from H. H. Munro (Saki), *The Story-Teller*.

Reprinted from Shirley Jackson, *The Lottery*.

Reprinted from Elizabeth Bowen, *Tears, Idle Tears*.

Reprinted from *All God's Chillen Had Wings*.

Reprinted from Ernest Hemingway, *Soldier's Home*.

Reprinted from D. H. Lawrence, *The Virgin and the Gipsy*.

Reprinted from Anton Chekhov, *The Kiss*.

Reprinted from Guy de Maupassant, *Little Louise Roque*.

Reprinted from O. Henry, *A Retrieved Reformation*.

Reproduced by permission from Michael Malone, *White Trash Noir*, © 2006.

Reproduced by permission from Lee Smith, *Desire on Domino Island*.

Reproduced by permission from Dorothy Weil, *Life, Sex, and Fast Pitch Soft Ball*, PublishAmerica, © 2005.

MORE UNZIPPED GUIDES

CONTENTS

INTRODUCTION

Shut Up and Write!

YOU WANT TO write a short story? You're out of your mind. Which, as it happens, is the best place for you to be. Start with UNZIPPING YOUR MIND. You have it all balled up in there. Intimidated. Angsty. Don't know where to start. Facing that blankety-blank-blank page. It's a bummer. Except it's not. It's fun, actually. You just need to learn a few tricks (and read a few short stories to see how they work).

How do you tell a story to your friends—relate a cool, funny, sad, hairy, edgy, nauseating, freaky incident? You do it out of your own head, in your own voice, in your own words, with your own curiosity and passion. Unzip all those tight threads and tell what went down, or what's going down, or what might go down.

When you're telling a story to a friend, your listener gets the point, because—though you may not realize it—the way to tell stories to people every day is this: *little story, point, little story, point, little story, final point.* You do it unconsciously. And in the end, your listener laughs, or cries, or is freaked out, or scared, or raving right along with you.

That's the way the hottest writers write, too. Sometimes, when you're reading, you don't even know you've gotten to the final point . . . and then . . . DOUBLE TAKE. WHAMMY. So *that's* what it's all about.

Come on, try it. UNZIP YOUR MIND. Read the following sentence, which makes perfectly good and logical sense. It may not seem to at first, but after you think about it awhile . . . DOUBLE TAKE. WHAMMY. Here's the sentence:

> *They say time flies, but you can't; they move at such irregular intervals.*

Um, say what? Read it again; think about it with your mind unzipped. It will finally come clear, when you're jarred out of your conventional thinking. Our heads hate that box of conventional thinking. That box is full of somebody else's words and thoughts. DON'T GO THERE! Such thinking keeps the lid on us, and on our original minds, and on our writing. Sometimes it keeps the lid on our good sense and fresh and creative approaches to issues, too. Come up for air.

Remember, writing OUT OF YOUR HEAD is better than any other writing you can do. When you're writing out of your head, you're more in control of your own thinking than you've ever been. Seem like a contradiction? Well, a lot of writing is just that. Writers think lofty and base thoughts, use nice-mannered words and street language, examine general stuff and specific stuff, let their deep, dark subconscious minds feed them instincts, and make startling, cool connections—and then writers become control freaks in revising. In the process, using YOUR SELF, everything that goes to make up *you,* produces the most intense writing.

Got the DOUBLE TAKE, WHAMMY yet? If not, you can use the cheat sheet on the very last page of this book to GET IT. A lot of people become reading cheats when they JUST CAN'T WAIT to see what happens in stories. That means the writer has written a good, engaging, suspenseful story that drives the reader on to want more. And you can do it, too. Keep on keeping on. It's SHORT STORIES. UNZIPPED.

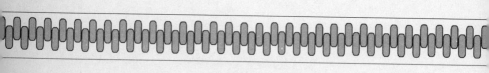

CHAPTER 1

So What's Wrong with a Couch Potato?

GOOD WRITER has to have certain qualities to begin with—or just *fogetaboutit.*

- *Curiosity* You need to say W*hat if?* and W*hy?* about a lot of things.

- S*tubbornness* You have to make yourself stick your butt in the chair and stay there, even when the writing is not going all that swell.

- *A Strong Imagination* How must it feel to be in that other person's place or to walk in his shoes?

- *Focus* Forget that your buddies are calling you to go out, all those e-mails are piling up in your inbox, your girlfriend is about to dump you, the world is a mess, the dog is howling to go out. Take all those things into your writing, BUT SIT THERE.

- *Patience* To get it right. To revise and revise until it's right. You're the person driving, so while you'll look at the map now and then, your mind will go on side trips and pull in anything to make the story work.

To write, you have to be both a far-out fantasizer and a control freak.

WORDS UNZIPPED

Juxtaposition: Placing two things together that don't normally go together. This often creates surprise or humor (pigs and flowers, for example).

Tension: Two or more things or people in opposition or disagreement (she wants to go, he doesn't; he likes short women, she's tall). Tension is also built in a story just before the climax.

Characters: The actors or people in a story; they can be old, young, teenaged, comic actors, dramatic actors, onlookers, and so on. Stories usually have a few principal characters and several secondary characters.

Writing takes all these qualities and one more critical one. To be a good writer, you have to be a reader first. So move away from that TV remote and sit down. Better, lie down. For writers, reading books is like athletes eating healthy meals and working out. It's the fuel that makes things run. So read everything you can get your hands on. Absorb it—you never know when you'll use it. The more stuff you can put in your head to mix with everything else that's already in there, well, it makes that imagination strong—just like healthy food strengthens an athlete's body.

BUT WAIT! It's not all reading and learning. Sometimes, it's like, Chill. Lighten Up. It's UNLEARNING, too.

So forget a lot of things you've been told, such as: know what you're going to say before you say it; long is better than short; big words impress people; always outline first; there's one right way to write a good story; daydreaming is a waste of time; curiosity killed the cat; it's not nice to eavesdrop; an idle mind is the devil's playground. Your mind is hardly ever idle; it's a creator's playground. Mostly it's taking things in, maybe putting things together, maybe connecting the dots. Your mind is busy, even when it's chilling.

YOU HAVE EYES

Use them. Look at wha's happenin'. Remember what you see and write it down. That's called *capturing visual images.* Making your writing visual—so your reader can *see* what you're saying because you paint such clear pictures—is one of the best ways to hold your reader's attention.

You know how a filmmaker sometimes puts a character or two in brighter colors, or brings him or her up close to the camera, or moves in on eyes or lips or hands, or films some people to stand out while putting the others in soft focus or a blur? You can do that in writing, too. You can make your story almost like a film by controlling what your reader zooms in on. Novelist Joseph Conrad said, "My task which I am trying to achieve is, by the power of the written word, to make

WORDS UNZIPPED

Plot: The action in a story or the story line; what's happening.

Protagonist: The main character in a story around whom everything else revolves; this character can be either male or female and of any age.

Antagonist: The person or force that opposes the main character and which the character has to overcome; this can be an outside force—a storm, a stalker—or an internal force, maybe indecision or conscience.

WORDS UNZIPPED

Scene Setup: The introduction of characters, setting, time period, geographic location; usually comes at the beginning of a story.

Dialogue: At least 2 people talking. Dialogue is also what characters say to each other in a story.

Rising Action: The place in a story where the plot begins to take off in a certain direction; often follows scene setup and dialogue.

Complication: The tense point in the story when the protagonist and antagonist confront each other and the plot thickens or becomes complicated.

you hear, to make you feel—it is, before all, to make you *see*. That—and no more, and it is everything." You can't make anyone see anything unless you see it yourself. Eyeball things. Be a super-spy. Be a nuisance.

YOU HAVE EARS, TOO

Be a secret snoop. Listen. Listen to the rhythms people use when they speak, the words different people use, different dialects. Be a gossip monger: Pay attention to gossip; after all, gossip is composed of little stories about people— some sweet, some funny, some snarky, some intrusive, and some abusive. Doesn't matter whether they are true or not, just listen. You create your own truth in your story. And you do it with what you've seen and heard.

HOOKING UP

Notice how putting two things together that don't normally go together (pigs and flowers, for example, or sheep and clocks, as you'll see in a story later) create humor and surprise. This is called *juxtaposition*. The more you put odd things together—even if it's a character giggling at a funeral or a toe ring on a corpse—the more delightful, colorful, and engaging your story will be.

And notice how two or more things or people in disagreement or opposition create *tension*. There's no story without tension! She wants to go, he doesn't. He likes short women, she's tall. The mountain climber is terrified of heights. Dad wants him to be a lawyer; he wants to be a forest ranger. The valedictorian was dumped by the college admissions officer. The Medal of Honor went to the wimpiest man. The army recruitment officer is a conscientious objector. The respected preacher breaks all Ten Commandments. See? You're getting curious about the *backstory* here—the story behind the story, the big *WHY*?

JOT, JOT, JOT

Carry a little notebook with you everywhere you go and record things you see, things you hear, things that make you smile, things that make you curious, things that are *incongruous* (they don't

WORDS UNZIPPED

Climax: The high point of a story when everything comes to a head or crisis. The climax creates a turning point in the story.

Falling Action: After the climax, a release of tension that leads to the conclusion of the story or a tying up of loose ends.

go together). Record quotations you like. Clip weird news items from the newspapers. Become a person on whom nothing is lost, as writer Henry James said. Use all these things when you write.

WHAT'S HAPPENING?

Good stories have a *scene setup* and *introduction of characters, dialogue, rising action, complication, tension, climax, falling action, final tying up of loose ends* or *loose ends left dangling that readers have to tie up for themselves.* This is the structure of stories, whether short stories or long novels, whether in stage or film scripts. Each writer tinkers with the structure in different ways for his own purpose, or audience, or dramatic vision. The same way a house has a basic way of being constructed so that it won't fall down, stories do, too. Make your story stand up.

FEAR & LOATHING

Stories have a *protagonist* (the central character around whom other characters and action revolve). They also have an *antagonist* (the person or force that opposes the protagonist). The antagonist doesn't have to be another person—it can be a natural force (a storm, a tornado, a landslide, quicksand, a wild animal, a mountain the protagonist is determined to climb or a channel the protagonist is determined to swim), a force within the protagonist's own mind that he fights against, or a dilemma that divides her when she needs to act. The antagonist can even be a child, or children, or sometimes simply A FATE that the protagonist has to face, like a deadly disease, or a crippling accident, or the fact that someone he loves doesn't love him back.

Look for all this stuff as you read the stories here—and the stories you're going to read when you become a *real* couch potato—without a remote in your hand. Good stories will take you on better, richer, more exciting adventures than any reality TV show. Lie down and read! UNLEARN! Use all your senses all the time! You'll write better. (And you'll be smarter, too.)

CHAPTER 2
Saki Is More Than a Drink

TO HELP YOU start thinking about how short stories work, read the following little story. It's written by H. H. Munro, whose pen name was Saki. As you read, notice how the story unfolds: scene setup, introduction of characters, characters' personalities are revealed by what they say to each other (*dialogue*). Watch how our interest is gradually captured.

The Story-Teller

It was a hot afternoon, and the railway carriage was correspondingly sultry, and the next stop was at Templecombe, nearly an hour ahead. The occupants of the carriage were a small girl, and a smaller girl, and a small boy. An aunt belonging to the children occupied one corner seat, and the further corner seat on the opposite side was occupied by a bachelor who was a stranger to the party, but the small girls and small boy emphatically occupied the compartment. Both the aunt and the children were conversational in a limited persistent way, reminding one of the attentions of a housefly that refused to be discouraged. Most of the aunt's remarks seemed to begin with "Don't" and nearly all of the children's remarks began with "Why?" The bachelor said nothing out loud.

"Don't, Cyril, don't," exclaimed the aunt, as the small boy began smacking the cushions of the seat, producing a cloud of dust at each blow.

"Come and look out of the window," she added.

The child moved reluctantly to the window. "Why are those sheep being driven out of that field?" he asked.

"I expect they are being driven to another field where there is more grass," said the aunt weakly.

"But there is lots of grass in that field," protested the boy; *"there's nothing else but grass there. Aunt, there's lots of grass in that field."*

"Perhaps the grass in the other field is better," suggested the aunt fatuously.

"Why is the grass in the other field better?" persisted Cyril.

The frown on the bachelor's face was deepening to a scowl. He was a hard, unsympathetic man, the aunt decided in her mind. She was utterly unable to come to any satisfactory decision about the grass in the other field.

The smaller girl created a diversion by beginning to recite "On the Road to Mandalay." She only knew the first line, but she put her limited knowledge to the fullest possible use. She repeated the line over and over again in a dreamy but resolute and very audible voice; it seemed to the bachelor as though some one had had a bet with her that she could not repeat the line aloud two thousand times without stopping. Whoever it was who had made the wager was likely to lose his bet.

"Come over here and listen to a story," said the aunt, when the bachelor had looked twice at her and once at the communication cord.

The children moved listlessly towards the aunt's end of the carriage. Evidently her reputation as a story-teller did not rank high in their estimation.

In a low, confidential voice, interrupted at frequent intervals by loud, petulant questions from her listeners, she began an un-enterprising and deplorably uninteresting story about a little girl who was good, and made friends with everyone on account of her goodness, and was finally saved from a mad bull by a number of rescuers who admired her moral character.

"Wouldn't they have saved her if she hadn't been good?" demanded the bigger of the small girls. It was exactly the question that the bachelor had wanted to ask.

"Well, yes," admitted the aunt lamely, "but I don't think they would have run quite so fast to her help if they had not liked her so much."

"It's the stupidest story I've ever heard," said the bigger of the small girls, with immense conviction.

"I didn't listen after the first bit, it was so stupid," said Cyril.

The smaller girl made no actual comment on the story, but she had long ago recommended a murmured repetition of her favorite line.

"You don't seem to be a success as a story-teller," said the bachelor suddenly from his corner.

The aunt bristled in instant defense at this unexpected attack.

"It's a very difficult thing to tell stories that children can both understand and appreciate," she said stiffly.

"I don't agree with you," said the bachelor. "There was a little girl called Bertha, who was extraordinarily good."

The children's momentarily-aroused interest began at once to flicker; all stories seemed dreadfully alike, no matter who told them.

"She did all that she was told, she was always truthful, she kept her clothes clean, ate milk puddings as though they were jam tarts, learned her lessons perfectly, and was polite in her manners."

"Was she pretty?" asked the bigger of the small girls.

"Not as pretty as any of you," said the bachelor, "but she was horribly good."

There was a wave of reaction in favor of the story; the word horrible in connection with goodness was a novelty that commended itself. It seemed to introduce a ring of truth that was absent from the aunt's tales of infant life.

"She was so good," continued the bachelor, "that she won several medals for goodness, which she always wore, pinned to her dress.

There was a medal for obedience, another medal for punctuality, and a third for good behavior. They were large metal medals and they clicked against one another as she walked. No other child in the town where she lived had as many as three medals, so everybody knew that she must be the extra good child."

"Horribly good," quoted Cyril.

"Everybody talked about her goodness, and the Prince of the country got to hear about it, and he said that as she was so very good she might be allowed once a week to walk in his park, which was just outside the town. It was a beautiful park and no children were ever allowed in it, so it was a great honor for Bertha to be allowed to go there."

"Were there any sheep in the park?" demanded Cyril.

"No," said the bachelor, "there were no sheep."

"Why weren't there any sheep?" came the inevitable question arising out of that answer.

The aunt permitted herself a smile, which might almost have been described as a grin.

"There were no sheep in the park," said the bachelor, "because the Prince's mother had once had a dream that her son would either be killed by a sheep or else by a clock falling on him. For that reason, the Prince never kept a sheep in his park or a clock in his palace."

The aunt repressed a gasp of admiration.

"Was the Prince killed by a sheep or a clock?" asked Cyril.

"He is still alive, so we can't tell whether the dream will come true," said the bachelor unconcernedly; "anyway, there were no sheep in the park, but there were lots of little pigs running all over the place."

"What color were they?"

"Black with white faces, white with black spots, black all over, grey with white patches, and some were white all over."

The story-teller paused to let a full idea of the park's treasures sink into the children's imaginations; then he resumed.

"Bertha was rather sorry to find that there were no flowers in the park. She had promised her aunts, with tears in her eyes, that she would not pick any of the kind Prince's flowers, and she had meant to keep her promise, so of course it made her feel silly to find there were no flowers to pick."

"Why weren't there any flowers?"

"Because the pigs had eaten them all," said the bachelor promptly. "The gardeners had told the Prince that you couldn't have pigs and flowers, so he decided to have pigs and no flowers."

There was a murmur of approval at the excellence of the Prince's decision; so many people would have decided the other way.

"There were lots of other delightful things in the park. There were ponds with gold and blue and green fish in them, and trees with beautiful parrots that said clever things at a moment's notice, and hummingbirds that hummed all the popular tunes of the day. Bertha walked up and down and enjoyed herself immensely, and thought to herself. 'If I were not so extraordinarily good, I should not have been allowed to come into this beautiful park and enjoy all that there is to be seen in it,' and her three medals clinked against one another as she walked and helped to remind her how very good she really was. Just then an enormous wolf came prowling into the park to see if it could catch a fat little pig for its supper."

"What color was it?" asked the children, amid an immediate quickening of interest.

"Mud-color all over, with a black tongue and pale grey eyes that gleamed with unspeakable ferocity. The first thing that it saw in the park was Bertha; her pinafore was so spotlessly white and clean that it could be seen from a great distance. Bertha saw the wolf and saw that it was stealing towards her, and she began to wish that she had never been allowed to come into the park. She ran as hard as she could, and the wolf came after her with huge leaps and bounds. She managed to reach a shrubbery of myrtle bushes, and she hid herself in one of the thickest of the bushes. The wolf came sniffing among the branches, its black tongue lolling out of its mouth, and its pale grey eyes glaring with rage. Bertha was terribly frightened, and thought to herself: 'If I had not been so extraordinarily good, I should have been safe in the town at the

moment.' However, the scent of the myrtle was so strong that the wolf could not sniff out where Bertha was hiding, and the bushes were so thick that he might have hunted about in them for a long time without catching sight of her, so he thought he might as well go off and catch a little pig instead. Bertha was trembling very much at having the wolf prowling and sniffing for her, and as she trembled, the medal for obedience clinked against the medals for good conduct and punctuality. The wolf was just moving away when he heard the sound of the medals clinking and stopped to listen; they clinked again in a bush quite near him. He dashed into the brush, his pale grey eyes gleaming with ferocity and triumph, and dragged Bertha out and devoured her to the last morsel. All that was left of her were her shoes, bits of clothing, and the three medals for goodness.

"Were any of the little pigs killed?"

"No, they all escaped."

"The story began badly," said the smaller of the small girls, "but it had a beautiful ending."

"It is the only beautiful story I have ever heard," said Cyril.

A dissentient opinion came from that aunt.

"A most improper story to tell young children! You have undermined the effect of years of careful teaching."

"At any rate," said the bachelor, collecting his belongings preparatory to leaving the carriage. "I kept them quiet for ten minutes, which was more than you were able to do."

"Unhappy woman!" he observed to himself as he walked down the platform of Templecombe station; "for the next six months or so those children will assail her in public with demands for an improper story!"

MORE TO UNZIP
POOR BERTHA, SHE'S HISTORY

How many short stories do you find that truly have some flesh on the skeletons of the story? Why do the kids like the man's story better than the aunt's? What elements does the man's story have that the aunt's does not? The aunt's story has an obvious moral. Does the man's story have a moral? What brought on Bertha's doom? Are stories obliged to be "proper" or have a moral? What makes the story amusing?

CHAPTER 3

Get with the Mood, Dude

WHAT MAKES A good story? A good story has everything life does: characters, action, suspense, drama, usually a twist of some kind, a beginning, a middle, and an end. Saki's story is set up early. He shows us where the characters are and he shows us who the characters are—3 children and their aunt and a stranger who seems to be annoyed by the kids. (The boy is the only one who has a name. Do you want to speculate on why? Does it matter?) The kids are restless and the aunt wants to keep them entertained. So she decides to tell them a story. The story is boring and moralistic. The kids have heard this story a hundred times before; it's a story with an obvious lesson. The story lacks suspense, drama, a twist at the end. Who wants to hear this kind of story? The man decides to take over to keep the kids quiet. His story has color, specifics, suspense, drama.

EXTREME MAKEOVER

Let's look at how he changes the aunt's story. Notice that he gives the little girl a name—Bertha. This allows the kids to identify with her and begin to visualize Bertha as a real person, just like themselves. He makes Bertha very, very good, as the aunt did, but when the kids' interest begins to flag, he adds just exactly *how* she was good—even to eating milk puddings as if they were jam tarts. His story leaves room for questions, for audience involvement. *"Was she pretty?"* one of the girls asks. Again, he includes the children with his answer, *"Not as pretty as any of you."* Then he adds, *"She was horribly good."* The kids leap on "horribly" in juxtaposition with "good." This arouses their curiosity. Surprise! A real grown-up can put these two words together in a way children understand. This is not going to be another preachy story. It has truth in it and some tension they can understand. Nobody likes extraordinarily good little anythings! That is not like life.

Then the man goes on to tell about Bertha's three medals for her great goodness, which she always wears pinned to her dress. Now the children have two reasons to dislike Bertha: she is not only good, good, good, but she wears her goodness on her clothes, her outside, to flaunt it to the world. This is enough to make the kids throw up. And notice that the man tells the children exactly what these medals are for: obedience, punctuality, and good behavior—not just some general goodness. Cyril, the little boy, repeats the words *"horribly good,"* because that's the way he thinks about extra good children and their behavior.

Well, then the resident Prince has heard that Bertha is so good that he lets her walk in his special park. But Cyril doesn't think this park is special because it has no sheep. He is, by this time deeply involved in the story and he wants to know more details. At this point, the aunt is feeling smug (*smiles and grins*) because her lame answers to Cyril's earlier questions have involved sheep. Then the man surprises all of them. He tells them about the mother having a dream that a sheep has killed her son, the Prince, or else a clock has fallen on him and killed him.

This is quite a surprise. Sheep have nothing to do with clocks. This is another startling opposition of specific things used to create tension. Even the aunt becomes admiring of the man's storytelling talents.

Then we learn that there are lots of little pigs running around in the garden. Well, you may remember that one of Bertha's good behaviors is keeping her white pinafore clean. Which animal comes to mind as the dirtiest animal in the world? So here we have pigs in Bertha's extra good, pristine, clean park. Another juxtaposition of opposites that creates humor and surprise. The man goes on to describe the pigs very specifically, with lots of color, black, white, spotted, and so on, so the kids can visualize them.

Notice that the story-teller then pauses to let all this sink in. This is called *pacing*. A good story-teller paces his story, using pauses, some down time, and then acceleration, moving briskly forward. Keeping the momentum going, with appropriate pacing, is a trick of good

WORDS UNZIPPED

Pacing: The timing of the action in a story or in the telling of a story; pausing then accelerating or maybe comic relief just after a dramatic scene.

story-tellers. Stand-up comedians and people who tell jokes people laugh at know this, too.

Then the story-teller gives us another opposition: pigs and flowers. The Prince has chosen pigs over flowers. Even the children recognize that this is an odd choice, and it delights them.

The man continues with a lot more specifics about the park: colorful, visual, and surprising. Ponds with colored fish, trees with talking parrots, hummingbirds that hum songs they recognize. Bertha can hardly enjoy all this because she is constantly thinking of how good she is, and how good she must remain (not even any flowers to *not* pick, oh phooey!). This park is like Eden, and Bertha is disappointed that she can't show her great goodness yet again. This park throws her off—she wants to be tempted to be bad so she can show how morally superior she is.

Then what happens? Not a snake to tempt her, but a wolf to stalk her, attracted by her white, white dress. But Bertha thinks she outwits him when she dives into the bushes. This ferocious wolf with the lolling black tongue has been foiled by a Miss Priss. But oh, what does the wolf hear? The clanking of Bertha's big, ostentatious medals. This snarling, hideous wolf had been out to eat some dirty little pigs, but now whom does he devour to the last morsel? Poor clean Bertha. She gets her comeuppance for her showing off and her pride in goodness.

Notice how the kids ask whether any of the little pigs were killed. The pigs are of more interest to the kids than goody-goody Bertha! Well, whew. The pigs all escaped. That's what the kids have identified with.

Notice that the kids think the story is *beautiful* but the aunt sniffs and says it's improper and doesn't teach the correct lesson to the kids. The man understands that he has been able to keep the children quiet, when the aunt, with her wimpy story, could not. As he leaves the train, he is delighted that he has told a story that will prompt the kids to tease their unimaginative aunt to come up with "improper" stories.

MORE TO UNZIP

NAME IT, IT STANDS OUT

Did you come to any conclusion about why only Cyril has a name? Maybe Saki wants us to focus on this one child by developing his character more than the others, by making him stand out as the principal *antagonist* of his aunt (the *protagonist)*, while the girls are just a hazy nuisance (like the blurred background people that a filmmaker sometimes produces). Also, by not giving any other characters names, maybe the author wants to *universalize* the story—these could be any kids, any aunt, any man. These people could be you, us. Good writers/story-tellers manipulate their material—and their readers—that way. They are in perfect control of all their elements. They can let you in (involve you and your own imagination) or hold you at a distance.

What is the final twist here? The man, a bachelor who presumably doesn't even have any kids and is annoyed by these kids in the carriage, is the story-teller of the title, not the fussy aunt, as we are first led to believe.

A simple story, but one that teaches story-tellers, as well as children, a lesson. What is the kids' lesson? It's a much deeper moral lesson than the aunt's story, which boringly preaches that children should be good. The man's story lets the kids understand in a charming, suspenseful, fanciful, dramatic way, that people with good behavior don't go around showing it off. People with in-your-face smugness get eaten by wolves!

CHAPTER 4

Unzip Your Imagination

EACH CHRISTMAS, A huge department store in New York City fastened pictures, in footprint shapes, to the floor running from the main door, across the room, up several flights of stairs, to lead eventually to the joys of Toyland, where children and parents could browse and shop. You couldn't miss them. They would lead you directly to an adventure of many delights. It was fun to watch the faces of the children as they stepped carefully in anticipation of what they would find.

Writers work that way, too. They virtually take you by the hand and lead you through a story, but you have to follow all the footsteps they lay down to reach the payoff. The more carefully you pay attention to every detail, the closer you will come to fully understanding the story. Some stories are quite simple on the surface, as Saki's is. Yet, as we've just seen, a lot of rather complicated stuff is going on—if you follow the footsteps alertly enough to see the pattern the author writes for you. The more skilled the writer, and the reader, the bigger the reward. That's why many of us get hooked on books very early and then keep following those footsteps the rest of our lives. And writers get hooked on writing books—a game and a challenge. And who doesn't love a game when you get to make up the rules and nudge other people to follow them?

WORDS UNZIPPED

Genre: Category of story; genres are mystery, romance, horror story, family story, and so on.

WHAT'S IN A GENRE?

As we've noted, good writers are control freaks, but they're also filled with fantasies. Fantasies are stories we like to escape into. Saki's story within a story is a *fairy tale* intended for children, but his *real* story is a richly illustrated and disguised bit of *propaganda*—well-thought-out, calculated instructions for story-telling adults. Let's take a look at some other genres of stories now.

Mysteries/Suspense Stories

These are the most intellectual of short stories because there is a puzzle at the core that the reader wants to solve, along with the sleuth of the story. People who enjoy mysteries tend to enjoy crossword puzzles and long novels such as Daphne du Maurier's *Rebecca,* or Bronte's *Jane Eyre,* or old movies like *Gaslight,* where something creepy and weird is going on, but you're not sure what. Mysteries must be carefully plotted—the footprints to the solution must be craftily and logically set down for the reader to follow. But they must not lead the reader to the solution too soon; nor should they be so haphazardly laid down that the reader just gets confused and can't follow them at all.

Family Tales

Basically, these tales are just that—stories that revolve around a family or families and reveal the different personalities, relationships, deep love, and rivalry within and among family members. In these short stories, the action, both physical and psychological, is fairly limited to one or two incidents or events and takes place in a limited period of time.

Romances

Boy meets girl. Well, yes. But that's hardly a story. What *happens* is the story—how do they meet? Where? Why is the meeting significant? Are they attracted to one another? Why or why not? Now we're getting to the story. What is the backstory of each? Have they been jilted, unhappily married, happily married? Has one, or both, been widowed? The story is not only what happens when and after they meet but also the baggage each brings to that meeting.

WORDS UNZIPPED

Narrator: The person who tells the story; this person can know everything that's going on or can have a limited knowledge of characters and events in the story.

Subject: The content or substance of a story; the main idea or event the author is writing about; a character can also be the subject of a story.

Horror Stories

We've all felt fear to some degree or another, so scary stories are the most immediately accessible to us—we identify with the emotion. Kids love to hear spooky stories around a campfire, where the surrounding darkness can make them feel creepy, but the presence of their buddies to huddle with, tingle with, and shiver with makes them feel safe. All of us, no matter what our age, like to tingle and shiver and get wide-eyed with excitement—as long as we can do so in safety. Horror, scary, spooky stories allow us to do that.

Stories with Surprise Endings

Here, the subject can be about most anything, and this story can be a family tale, mystery, adventure, romance, sketch—anything. The point is that it's short. It leads the reader on to one conclusion, but in the end, the writer slips in a surprise, which the reader never anticipated. It's almost the opposite of a mystery, where the reader knows all along that there's a puzzle to solve and has certain expectations. In the surprise-ending story, readers don't usually see it coming. But in writing this kind of story, you never announce or hint at the genre it is; the surprise is that the story-teller takes readers where they don't expect to go. O. Henry was a master of this genre.

Sketches

These are probably the easiest short pieces to write because they're not exactly entire stories; they don't always have a point—or at least not an obvious one. Sketches are little stories, scenes, or anecdotes, extremely limited in focus, but they can be *unlimited* in their subjects, their style, the voice they are told in, or their tone, which can be anything from hilariously funny to very dark and sober. Sketches are little slices of life. Writing sketches is a good way to remember incidents or observations that you don't want to forget, especially when you don't immediately see a short story evolving from them.

THE MAP

We've already talked about characters—protagonist and antagonist. Now, we want to talk about what happens in the story—the *action*—and the *voice*—the person who is telling the story.

Plotting

Who's doing what, to and with whom? Stories must have action, but the action must have meaning, leading to some insight into the character or interest by the reader. People running around, or shooting each other, or kissing each other is not a plot. The plot must be the context of what all this running, shooting, kissing means; what's the motivation, what's the goal? What's the WHY?

The Seesaw

In writing fiction, you need to decide whether you want *Character* or *Plot* to be your principal driving force. Plot-driven stories (sometimes found in mysteries and adventure stories) force the reader to be more interested in the Action than in the Characters. In Character-driven stories (often family stories and romances), the reader becomes more interested in the Character than in the Action.

Plot and Character sit on opposite ends of a seesaw—one can tip the other up or down or the two can hold a balance. The one who's holding the seesaw down is in control. Say Plot is down, holding Character up there, teetering on the high end. Plot is the focus, is in charge of what Character does, and is the dominant or driving force in the story. Or it can be that Character is holding the seesaw

MORE TO UNZIP

What If?

We all have our own personal set of fantasies we entertain ourselves with:

- Newfound riches *What if I won the gazillion-dollar lottery?*

- Newfound freedom *What if I could do anything I wanted to, anytime I wanted to?*

- Newfound love *What if the guy or girl of my dreams walked into the room right this minute and fell for me?*

Notice how our personal fantasies all begin with *What if*. *What if* is a good first step in writing a short story. *What if I took this particular girl or guy and placed her or him in this particular situation and watched what happened?*

down, so Plot is driven by what Character does. These two driving forces, Plot and Character, are in constant dynamic interaction—what each does affects the other.

Again, think of movies: chase movies, crash 'em up movies, action movies. Plot is doing that chasing, driving that car. We care about the Characters, of course, and hope they do not die (or do, depending on whether we think they're "good" or "bad" guys), but what absorbs us, excites us, is the chase, the Action. That's a Plot-driven story. But if we're more absorbed and excited by who the Character is, just how "good" or "bad" he or she is, what's in that Character's mind or backstory that the author has given us, and just want the chase to be over so that our favored guy survives, that's a Character-driven story.

The best stories are those in which we see Plot and Character hold each other in a balance. One teeters, the other totters, and then the dynamic changes. The two work together to drive the story.

WORDS UNZIPPED

Voice: How your narrator speaks. The words and language you choose to put in the narrator's mouth.

Tone: The way the voice speaks in a story. The tone is determined by the attitude the speaker takes toward the audience.

Audience: Who is the story for—children? adults? teenagers?

WORDS UNZIPPED

Attitude: The way the author or narrative voice takes the subject of a story. The voice telling the story can take the characters and plot seriously, humorously, neutrally, passionately, and so on, and that attitude is sensed by the reader in the tone the voice uses. (For example, a voice can have an angry tone, but a reader can sense in that tone an attitude of, perhaps, self-justification.)

What's the Point?

Someone with a point of view (a *narrator*) must tell your story. The writer may tell the story herself. Or he may let one of his characters tell it. He may tell the story through the thoughts and feelings of only one character or through two or more characters. The writer may choose to tell the story through letters or diaries that the writer herself reads and relates, or she may let a character read these letters and interpret them. The writer decides how much the story-teller knows about each character and the situation.

A writer can use variations or combinations of the following narrative points of view, but here are the basics.

First Person (major or minor character tells the story)

One person tells the story and does so by using I—*I saw this, I think that, my name is Isaiah.* This narrator brings us immediately into the story, as if the story is being told directly to us. This is a very flexible narrative voice because the person speaking can know everything that's going on in the story, have only a limited amount of information, flash back, flash forward, speculate, think aloud, analyze a situation, or overlook obvious things going on.

The author disappears into the character who tells the story. We know the story *only* from what this one person tells us, and we can be made to choose to believe that person or not. Just as in real life, some people who tell us something tell the truth, and some people's perceptions are unreliable, and some people just outright lie. But it's up to us to make those calls—with clues from the author.

Here's an excerpt from a short story by Michael Malone with a first-person narrator. Charmain is the main character (protagonist), as well as the narrator, and here she tells us about sitting as a defendant in a court of law, on trial for her life. This is the opening passage of the story. Notice how Malone gives her a fresh, engaging voice by manipulating the language—the way she speaks, the terms she uses, what details enter her mind, how she sees things. The story is:

White Trash Noir

All of a sudden Dr. Rothmann, the foreman of my jury, says she wants to talk to the judge. She gives me a look when she walks by the defendant's table, straight in my eyes, and I nod back at her but I can't tell what she's thinking because there's so many different feelings in her face. But behind me my Mawmaw stands up and bows her head to her. The judge and the jury get up too and they crowd each other out of the courtroom and just leave us sitting here.

My lawyer leans over and says, "Charmain, you have got to change your mind and take the stand." And I tell him, "No thank you."

Mr. Snow goes, "This is Murder One, Charmain. You just cannot kill your husband in the state of North Carolina if he played ACC basketball."

I go, "Well, this is Charmain Luby Markell and I'm not talking about my personal private life to a bunch of strangers in a court of law and have them turn it all into lies against me and mine."

I got this lawyer? He's young, just two years more than me, and halfway through our first talk in the jail I can tell he hasn't had a lot of Life Experience, which, between you and I, I've already had way too much . . .

So Mr. Snow wanted me to get up on the witness stand and tell why I shot my husband in the head and set him on fire in our backyard.

Mr. Snow chews at a cuticle; his nails are a mess. He sighs a long deep sigh and shakes his head at me. "Please won't you help me here, Charmain?"

Please won't I help him? Who're they trying to give a lethal injection to, me or Tilden Snow III? . . .

Note here that the author tells us a great deal about Charmain by letting her tell the story herself. We know what she sees and knows and what she doesn't know ("*I can't tell what she's thinking . . .*"). We know she has someone in court to support her ("*my Mawmaw*"—

presumably her grandmother) and her lawyer, Mr. Snow, who doesn't, to a world-weary Charmain, seem to have much *"Life Experience."*

We also know that Charmain is stubborn (she will not testify) and private (she won't talk to strangers about her personal life). And we know why she is on trial for her life—she shot her husband in the head and set him on fire in the backyard. We know that she is guilty from the get-go because she tells us herself, and, at this point anyway, we have no reason to disbelieve her.

All we know is filtered through Charmain's mind and her words. We are pretty shocked when she tells us right off that she killed her husband in a grisly way, but we still admire her honesty a little. The author makes us curious to see where this story is going, whether Charmain really did what she claims (and we sort of hope she didn't, because in these few lines she has made us, through her engaging voice, sympathize with her and we can't imagine how she can get off if she committed this hideous crime).

Also by her voice, we can guess at her age and level of education (*"I got this lawyer?"*). Young people, and especially young girls, often make statements that end in question marks, as if they are asking you to agree with their simple statement. Her level of education shows in the phrase *"between you and I"*—which is poor grammar, often used by poorly-educated people who want to be "correct" but get it wrong.

The title, *White Trash Noir*, also gives us a clue about the class of people we are going to meet in this story. *Noir* means *black*, but it's also a style or genre associated with movies of the 1940s and

'50s—shadowy, dark movies, usually set in large cities, revealing the seamy, underbelly of life (examples: *The Sweet Smell of Success, Double Indemnity, Laura*). Our author sets his story in a small town among "white trash," or what we usually think of as low-class people.

Finally, the author has let us know that this story is going to be somewhat comical (even though *noir* is hardly ever comical), in spite of its subject matter, with the early comment: *"This is Murder One, Charmain. You just cannot kill your husband in the state of North Carolina if he played ACC basketball."* And Charmain's comment—"Who're they trying to give a lethal injection to, me or Tilden Snow III?"—is also humorous and tells us that, though she may indeed be a criminal, Charmain is spirited and nobody's dummy.

Through the voice Malone selects to tell this story, and his manipulation of that voice, we already know a great deal about our protagonist here—and we like her (please don't give her a lethal injection!).

All of this is done by the author's skillful use of his narrator and the language she uses. This story is a MYSTERY that sort of works backward. We know who the criminal is and the crime (or think we do), but the puzzle is why did she do it? Or if she didn't, why would she confess? And what's going to happen to her? It has mystery, suspense, and drama—all created in a few lines of the opening passages.

This is a Character-driven story (we care more about Charmain than about the plot). Murder mysteries are not usually written from the viewpoint of the murderer because that voice may give away too much to the reader too early. An author interested in exploring a criminal's psychology might use this point of view, as Dostoevsky did in his novel, *Crime and Punishment,* illuminating the psychology and morality of a murderer. Examining those elements, in a more condensed way, is what Malone does in his story about Charmain.

Here's another delightful use of first-person narrator. As you read, try to count the things you learn about the speaker by just listening to her talk. The story is from a short novel (or *novella)* constructed like a long short story. It's called *Life, Sex, and Fast Pitch Soft Ball;* the author is Dorothy Weil. Here is the opening chapter:

Mall Pall

We're at the mall, right?

Me and my dad are spending quality time together.

Dad looks like Ken Griffey Junior 'cause a big picture of Junior is on the sports page Dad is hiding behind. We're sitting by the fountain where the old men and husbands waiting for their wives hang out. We haven't exchanged much conversation. On the way over Pop said, "Cover up your navel," and complained about how long he had to wait out in the car for me, considering how little I was wearing.

"What on earth do you women do?"

I shrug . . . Mom, a size six who actually looks like Barbie, changes clothes five times before she goes out the door. But me . . . it takes awhile to find something to wear since I've porked up another five pounds (age 14, height 5 foot 1, weight—well, never mind).

"Course I'd know what goes on in my own house," Pop says, "if I were allowed in it."

Nothing I can do about him coming in the house; it's OK by me, but the lawyers have worked it out that Mom is in control there and says who comes and goes. He's still hoping their split won't end up in divorce court.

I paw at my Dad's newspaper like a puppy, and he lowers it for a minute.

"How's school?" he asks, regular as clockwork.

"Boring."

He puts the paper back up. End of conversation. Nothing for me to do at a time like this but turn to the drug of choice, shopping.

Well, you know a good many things already about the narrator, don't you? But you don't know her name. In first-person narration, you'll learn the narrator's name when someone addresses her, or when she reads a note addressed to her, or when she sees her name posted somewhere and repeats it. Or when she tells us herself, as our narrator does here a few lines later:

Maybe I should get a tattoo, I think. If I do, what would it be? Cute with teddy bears or hearts? Or dirty or maybe environmental? Save the whales? Oh, yeah, I can just hear the little jokes now at school: joining a self-improvement group, are we? If I put my own name,

Mercedes, on my butt I'll get more laughs. Already kids call me "Mer-say-des" like the car and make a few size comparisons. (It's Mer' cedes). Maybe I'll just get a couple of rings in my nose.

This is a FAMILY TALE, as you could probably tell by reading these passages. But if you read the whole story, you'll see that it goes way beyond family doings. (Actually, Mercedes joins a fast-pitch softball team to lose weight—and learns a lot about herself and other people from playing on the team.)

Objective (narrator is outside the story and reports it)

Using a third-person-objective narrator, the author shifts the point of view outside the story itself to a teller, who calls the characters *they* or *them* or *he* or *she*. This narrator sees the characters go about their business and comments on them, a sort of silent, roving camera. This is the way Saki tells us the story of *The Story-Teller*. The author does not comment or interpret. We read the story as if we're watching a movie or play because we see only what the characters see and hear only what they say. We have to interpret what's going on by ourselves.

Here's a brief excerpt from Shirley Jackson's *The Lottery,* told from an objective point of view.

The Lottery

The morning of June 27th was clear and sunny, with the fresh warmth of a full-summer day; the flowers were blossoming profusely and the grass was richly green. The people of the village

began to gather in the square between the post office and the bank,
around ten o'clock; in some towns there were so many people that
the lottery took two days and had to be started on June 26th . . .

Notice that the objective narrator is simply setting the scene here in a calm, reporter-like way. The story, actually, is horrifying and grossly sick, but you wouldn't know that by the way the narrator is reporting it.

Omniscient/Limited Omniscient

A third-person narrator can know everything there is to know about the characters (*omniscient narrator/all knowing narrator*). Or this teller can know some things about the characters and story, but not others (*the limited omniscient narrator*).

Here is a brief bit from Elizabeth Bowen's *Tears, Idle Tears*. An omniscient narrator lets us into the thoughts and perceptions of Frederick, his mother, some observers, and almost everyone in the story. This narrator knows ALL and tells us.

Tears, Idle Tears

Frederick burst into tears in the middle of Regent's Park. His mother,
seeing what was about to happen, had cried: "Frederick, you can't
in the middle of Regent's Park!" . . . Frederick, knees trembling,
butted towards his mother a crimson convulsed face, as though he
had the idea of burying himself in her. She whipped out a hand-
kerchief and dabbed him . . . exclaiming meanwhile in fearful
mortification: "You really haven't got to be such a baby!" Her tone

attracted the notice of several people, who might otherwise have thought he was having something taken out of his eye . . .

The omniscient narrator here gets into the minds and feelings of everyone in the story.

The limited omniscient narrator can get into the minds, feelings, and perceptions of only one character, as below in this excerpt from:

Dark

Tuesday, 4 pm: Hank Jordan was in a dark mood. The world was too much with him. His wife, Rebecca, still at home recuperating with a broken leg from a fall in that pock-marked driveway he hadn't fixed. Mark, still struggling financially, and with those two kids. And Pete, floundering around trying to become an actor out in Hollywood. Why had his two sons never quite grown up, taken hold of their own lives by themselves? What had he done wrong?

Maybe it was him. Maybe it was his job—maybe a cop's job took too much time and energy. Maybe if he'd just made enough money to send them to better schools . . . Maybe it was just fate.

Fate. He pushed away from his desk, scooting his wooden chair back on the wooden floor, which made a groaning noise that matched his mood. A tall man, still fit and athletic in his forties, he wandered over to the communal coffee pot just outside his office door and poured another cup, thinking. He'd read Oedipus in college, and he figured that knowing your future in advance just led you right into it. No more control than an insect marching resolutely into a Roach Motel—you check in, you don't check out. You couldn't circumvent fate, and it was foolhardy to try. Fate was just how your life turned out for you, that's all . . .

Like for Jim Manchester. Hank sighed and pulled out the six-year-old yellowed file again. What was it about the case that

MORE TO UNZIP

SERIOUS, WITH SOME HUMOR ON THE SIDE, PLEASE

If you're writing a serious story, your voice takes on a serious, sober tone. But if the story is light and comical, the voice changes. Notice that even if the subject is serious, the tone can be light, as it is in Malone's story about Charmain on trial. After all, she's facing the death sentence, but the way she sees the trial has a humorous tone. And Hank, feeling down and guilty in the *Dark* story, has a serious tone.

kept eating at him? Jim, this young actor guy of twenty-four, had fallen down the back deck stairs of a photographer's house, crushed his head, splattered blood all over, and died. Not such a weird scenario, given all that cops see. But still something troubling here. It hadn't added up then, and it didn't add up now . . .

Notice that this narrator, though he or she stands completely outside of the action, tells the story from deeply inside the head of Hank. Here the narrator, and therefore the reader, knows *only* what Hank thinks and feels, not what others in the story may think and feel. We know Hank is sad and worried about some things and we can tell he also feels guilty about something.

This story combines a MYSTERY or SUSPENSE STORY with a FAMILY TALE. Hank feels guilty about his family, and he also feels guilty about not solving an old case of a man's death. Could it have been an accident or was it murder? Already, the reader is wondering.

CHAPTER 5

Introducing: Skeletons

IN THE FOLLOWING chapters, you'll hear bones rattling. These will be the skeletons of several different kinds of short stories for you to use as models upon which to hang the flesh of your own characters and plot.

THE HIP BONE IS CONNECTED TO THE THIGH BONE

Writing is thinking. Since this is true, it's usually limiting to make an outline before you begin writing your story. Often making a skeleton, or a brief plan of where you want to take your story, is best at this point. As you think and write, your story will begin to unfold. YOUR UNZIPPED MIND WILL TAKE YOU THERE. And after you've written a little—captured your characters, set them into a location, put them into action, made them talk to one another, and started to let your fantasy unzip in your imagination— AFTER all this, you may then want to write an outline or plan to finish your story.

FINDING THAT SKELETON KEY

Let's start with writing a *sketch* because sketches are brief, have no obvious point, have only a few characters, and are often told in first person. They're like telling an anecdote about something you've seen, heard, done, been a part of. They can even, like a joke, have a punch line (but often they don't). Sometimes, whatever point or insight is in a sketch is buried inside it and the reader has to find it (as in Saki's short story, though that has more flesh on its bones than a sketch).

Here's a little sketch that was originally a *folktale*—a story handed down orally through generations, often by people who did not read or write, such as early African Americans and Native Americans. Today, we call that kind of story *magical realism*. That is, there is down-to-earth realism, real people going about their real lives, but their authors give their characters magical powers that sometimes (as in the following story) give them hope and raise their spirits. (Remember the movie, *E.T.?* The little boy rides his bicycle toward the moon? That was also *magical realism.* Remember how you felt when his bicycle wheels lifted up into the air? Did your breath catch and did your spirit soar?)

The following sketch was passed on by Caesar Grant, a laborer from Johns Island, South Carolina, who said he heard it from his grandfather, who had heard it from a 90-year-old man. Here, an omniscient narrator gives us this story, which was later written down and collected in books on Negro Folklore. It's called:

All God's Chillen Had Wings

Once all Africans could fly like birds, but . . . their wings were taken away. There remained, here and there, in the sea islands . . . some who had been overlooked, and had retained the power of flight, though they looked like other men.

There was a cruel master on one of the sea islands, who worked his people till they died. When they died, he bought others to take their places. These also he killed with overwork in the burning summer sun, through the middle hours of the day, although this was against the law.

One day, when all the worn-out Negroes were dead of overwork, he bought . . . a company of native Africans just brought into the country, and put them to work in the cotton field.

He drove them hard. They went to work at sunrise and did not stop until dark. They were driven with unsparing harshness all day long, men, women, and children. There was no pause for rest during the unendurable heat of the midsummer noon, though trees were plenty and near. But through the hardest hours, when fair plantations gave their Negroes rest, this man's driver pushed the work along without a moment's stop for breath, until all grew weak with heat and thirst.

There was among them one young woman who had lately borne a child. It was her first; she had not fully recovered . . . and should not have been sent to the field until her strength had come back. She had her child with her, as the other women had, astraddle on her hip, or piggyback.

The baby cried. She spoke to quiet it. The driver could not understand her words. She took her breast with her hand . . . that the child might suck and be content. Then she went back to chopping knot-grass, but being very weak and sick with the great heat, she stumbled, slipped and fell.

The driver struck her with his lash until she rose and staggered on.

She spoke to an old man near her, the oldest man of them all, tall and strong, with a forked beard. He replied; but the driver could not understand what they said; their talk was strange to him.

She returned to work, but in a little while she fell again. Again the driver lashed her until she got to her feet. Again she spoke to the old man. But he said: "Not yet, daughter; not yet." So she went on working though she was very ill.

Soon she stumbled and fell again. But when the driver came running with his lash to drive her on . . . she turned to the old man and asked: "Is it time yet, daddy?" He answered: "Yes, daughter; the time has come. Go; and peace be with you!" . . . and stretched out his arms toward her . . . so.

With that she leaped straight up into the air and was gone like a bird, flying over field and wood.

The driver and overseer ran after her as far as the edge of the field, but she was gone, high over their heads, over the fence, and over the top of the woods, gone, with her baby astraddle of her hip . . .

Then the driver hurried the rest to make up for her loss; and the sun was very hot indeed. So hot that soon a man fell down. The overseer himself lashed him to his feet . . . The old man called out to him in an unknown tongue . . . and he leaped up in the air and was gone like a gull, flying over field and wood.

Soon another man fell. The driver lashed him. He turned to the old man. The old man cried out to him and stretched out his arms as he done for the other two, and he, like them, leaped up and was gone through the air, flying like a bird over field and wood.

Then the overseer cried to the driver, and the master cried to them both: "Beat the old devil! He is the doer!"

The overseer and the driver ran at the old man with lashes ready, and the master ran too, with a picket pulled from the fence, to beat the life out of the old man who had made those Negroes fly.

But the old man laughed in their faces and said something loudly to all the Negroes in the field . . . And as he spoke to them they all remembered what they had forgotten, and recalled the power which once had been theirs. Then all the Negroes stood up together, the old man raised his hands, and they all leaped up into the air with a great shout; and in a moment were gone, flying, like a flock of crows, over the field, over the fence, and over the top of the wood; and behind them flew the old man.

The men were clapping their hands; and the women went singing, and those who had children gave them their breasts; and the children laughed and sucked as their mothers flew . . .

The master, the overseer, and the driver looked after them as they flew, beyond the wood, beyond the river, miles on miles, until they passed beyond the last rim of the world and disappeared in the sky like a handful of leaves. They were never seen again . . .

So, how would you interpret this sketch? Did those slaves really fly—or did they all die together and send their souls soaring? Do you think the old bearded man might represent God, releasing them from their world of woe to fly away to heaven?

The story-teller just tells the tale, and we are left to make of it what we will. No moral. No real point. But a very moving, elevating little story.

Sketches can be about anything that makes you curious. Writer Ann Hodgman wondered what dog food tastes like, so tried a Gaines burger and decided to spend a week eating dog food—everything from doggie treats to the truly ikky, smelly, stringy, watery stuff in cans. Then she wrote a first-person sketch about her reaction to her experience in a piece called "No Wonder They Call Me a Bitch." Her voice, tone, and word choices make you feel as if you're right there eating the stuff along with her.

Here's another little sketch with a very different tone—not disgusting, sad, or sorrowful, or even finally uplifting, as the one about the slaves is. The tone here is humorous and rather outrageous—meaning over-the-top. Here the author recalls a real-life happening.

Dinner sans *Salad*

I am in one of those convenience cum food cum everything stores in New York City. My husband is outside wandering around, waiting for me. A foreign-born man is ahead of me at the counter, but the proprietor is somewhere in the back.

The man says something to me in a vaguely Germanic language, as he holds several rolls of breath mints toward me. I don't understand what he is saying at all, smile, and shake my head. He continues trolling the counters, looking at various breath mints, gums, and mouth washes. Clearly, I think, this man is paranoid about the pungency of his oral cavity.

I do my shopping: endive, avocado, arugula, and other healthy stuff for a salad. When I return to the counter, the man is still standing, looking puzzled. He is hunched over a tubular-shaped cartridge in his hand, trying to read the instructions on its side.

He looks up and hands it to me. I read the instructions and blanch, shake my head, say "no, no," and hand it back to him. He continues to stare at it.

He pulls off the cover and holds it up to his face. Opening his mouth wide, he simulates spraying it into his mouth. I grab his arm. "No, No! Nyet! Nyet! Non, Non, Pas, Pas, Ne Ne, Nul Nul. . . . Don't!" I yell at him, shouting through every variety of the negative I can muster, while shaking my palms in his face.

What I have read on the side of the tube is that the substance in it is Mace, sold on the streets of New York City, during a particularly menacing street-gang period. Looking fierce, I grab it from the man's hand, and he begins grabbing back. "No!" I yell. "Don't put it in your mouth? Dangerous! Dangerous!! VERBOTEN!!"

He keeps hold of the tube and wrestles me for it just as the shop-keeper appears. "What's going on here?" he yells.

"This man has mace!" I yell back. Thinking the man is trying to hold up the store, the proprietor tackles him. We are all scrambled on the floor yelling, when my husband appears, ambling in his chinos, delighted with the world.

"What???? What's going on?" He pulls me from the tangled threesome and attempts to set me on my feet.

"Too hard to explain. Just get me out of here," I wheeze, abandoning my healthy vegetables.

Outside, my patient husband, who claims I can have an adventure taking the garbage out to the curb, looks at me with quizzical, if not guarded, eyes.

"Don't ask. I'll explain over dinner—and, mind you, we won't be having salad this evening."

This sketch is based on a true incident that happened to yours truly. She thought it was a weird and funny incident, but it didn't contain enough material to base a short story on, so she wrote a sketch about it so she wouldn't forget it. The scene may turn up in a story or novel some day.

Remember that notebook you're supposed to carry around and jot in? Jot in it enough and you will have a gold mine of material. Flesh out your notes a little and you have a sketch. Fleshed out even more, a sketch may later turn into a short story.

Try taking a newspaper article that strikes you as weird or funny and turning it into a sketch. Like the one in this writer's files about the 90-year-old woman who died and left 2 million dollars in her will to her seven cats. Now, what can cats do with 2 million dollars? Was the old lady just dotty? Where did she get that kind of money, anyway?

Or how about the news clip about the 6-year-old boy who can chew cheese slices into the shape of every state in the Union, leaving a tooth mark where each capital city is located? Yours truly pulled it out of her files and used it in a novel several years after the news item appeared.

Or this one: An item about the odd names upwardly-mobile New Yorkers and celebrities give their babies. You probably have heard of actress Gwyneth Paltrow's baby girl, Apple, and Angelina and

Brad's baby, Shiloh. But did you know that *Today* host Matt Lauer and his wife named their third son Thjis (this, we are told, is pronounced Tice)? Other examples are Mitra, Sirus (pronounced Ziroos), Strawberry, Milo, Eowyn, Atlas, Zoos—and a character on the TV show *30 Rock* said she thought she would name her baby girl Bookcase. WHAT WERE THEY THINKING? Surely, there's a sketch in here somewhere. Maybe put all these kids on a playground and have them duke it out because nobody pronounces their names right!

There's enough material in these little tidbits to base a sketch on—if you, the writer, unzip your imagination and create all the details.

Though sketches are the easiest of all short pieces to write, they must be carefully shaped to keep them brief. Not a word must be wasted, and they must be written with carefully selected details and colorful words and images—just enough to make the little story vivid, but not enough to make it ramble. Sketches are usually tightly focused on one idea or incident.

In the last sketch, why do you think the author uses a French word, *sans* (*without*), in the title, and begins with a line including two *cum*s (Latin for *with*)? Can you guess?

Well, here's why. There wouldn't be any story to tell if there had not been a major lack of communication. And what caused that problem? A foreign language unfamiliar to the narrator and one unfamiliar to the man with the breath fixation. It's a tiny theme in the sketch and maybe the only point of the sketch. It's a very condensed concept—*with/without communication*—and the author looked for a condensed way of getting it across.

And why do you think the writer does not repeat the *sans* when speaking to her husband at the end? Well, that would have been overkill; besides she had no reason to speak French to her husband, who speaks perfectly good English. And the writer's point had already been made inside the store with people yelling at each other in different languages and nobody understanding anyone. Even the narrator didn't make herself clear when she yelled, *"He has mace!"* to the store owner. Everyone misunderstood everyone.

THREE IS CHARMED

It may take two to tango, but it usually takes three to make a short story dance. You *can* write a short story with only two characters (and playwrights often write two-character plays), but the dynamic between two characters is this: A's relationship to B; B's relationship to A. This is severely limiting. Take a romance story for example: A loves B, and B loves or doesn't love A. That's about as far as you can get with the story.

But add a third character and look at how the dynamics explode: A's relationship to B; B's relationship to A; A's relationship to C; B's relationship to C; C's relationship to A; C's relationship to B. That's a lot of relationships. If you add a fourth character, D, the dynamic possibilities explode to twelve, and so on the more characters you add.

Start with three major characters. You can add other, secondary characters, with their own little stories within the major story. But try making yourself stick to three major characters at first. Remember Saki's story? Five characters—3 kids, aunt, stranger. But the aunt and the stranger had a relationship to the kids *as a group*, not as three individuals. And the two little girls were *recessive*

MORE TO UNZIP

LET ME SEE WHAT YOU MEAN

Draw a triangle and label the angles A, B, and C.

Now, let's say you want to add D, E, and F (one letter on each side). Will you add D between angles A and B or should that be E? How should they all interrelate?

characters, while the little boy was the spokesman for them. So actually the story contained only three characters or six dynamic relationships. All clear?

CHAPTER 6

Skeletons in the Closet:
Mysteries & Suspense Stories

A **MYSTERY HAS** a secret or secrets. Somebody wants to know that secret and, in the end, somebody will. A crime is committed. Someone did it. Someone else wants to solve it.

BLOOD, HAIR, AND BONES: GOODS THIS SHORT STORY NEEDS

So, what are the crucial elements of a mystery? A crime, a criminal, a sleuth (amateur sleuth, professional cop, detective, private eye),

ZIP TIPS

All the Best Sleuths Have . . .

Curiosity, intelligence, courage, wit, intuition, keen observation and sharp senses, willingness to risk, shrewdness, nosiness, motive for getting involved. Sleuths must also be likable, usually quirky or with some fault with which the reader can identify. A sleuth can be a victim of the crime herself and her motive may be revenge. Or he may be accused of the crime himself and his motive may be exonerating himself. Sleuths may have any reason, but there *must* be a reason for getting involved.

witnesses, suspects, a motive, red-herrings (people the reader may suspect because they have a motive but no alibi), chase or entrapment scene. The sleuth's life or mental health must be at risk. Then sleuth outwits criminal, solves crime leading to a confession or arrest, and then loose ends are all tied up.

All along, the writer must keep the reader charged with curiosity about who the criminal is and the criminal's motivation. And *pacing* is vital here: You must not let the reader solve the crime too soon. But you must plant the clues all along the way so skillfully and so casually that when the reader gets to the end he will slap his head and say, "Oh, I should have known that!" If the reader speeds through the story and overlooks the clues right under his nose, then you know you have built up the suspense.

Mystery stories are fun to write because they tax *your* brain, as well as the reader's. And here is an example of why it's almost futile to try to outline your mystery story in advance. The following is an excerpt from a mystery novel by the author of this book. This is a stalking scene from the middle of the book. The first-person narrator/protagonist is Charm Hope, a college student, who, with her partner, Jake, is renovating an eighteenth-century Victorian carriage house on a pond on Long Island while they live in it. Jake is away on business. Gil is a police investigator to whom Charm has given information about a death she suspects is a crime. Here's the excerpt from:

The Bluebook Solution

Just then I hear the wind get up, banging the shutters, and I think of the lawn furniture likely to end up in the pond if I don't go pull it into the solarium. At the back door, I flip on the outside light. Shoot! Out again. The thing's always out. I dash to the edge of the lawn where Jake and I had set up a dark green table and four chairs we'd bought at an estate sale in Connecticut.

I drag two chairs back to the house just as the rain hits, warm and pelting, and am dragging a couple more toward the door, when I see a figure move out from the shadow of an oak tree. I can't tell in the dim light of the moon, diluted even further by heaving rain, whether the figure is male or female, but clearly it is following me toward the house. I don't know how, but you can always tell when someone is around who shouldn't be. Like a dog knows when to bark, I know when to run. I drop the chairs and run, gasping, my Doc Martens slurping in the mud, my heart drumming like it wants to leap out of my chest and beat the stalker to a pulp.

By feel, I find the door, get myself inside, slam and lock it. Who is that eerie person lurking outside? Nobody ever comes out this far at night.

I head for my room to dial 911. Then, I think, No, Gil. Dial Gil! As I rummage in my book bag for the phone, I hear feet on the basement stairs. Damn! The little door outside that leads to the basement must be open, and we never lock the door at the top of the stairs. My hands fasten around the phone and my eye catches the bathroom door, standing open a crack. I hear the stairway door creak open and footsteps coming down the long, dark hall. Only my room is lighted, so I know the intruder will head for that—and me.

I inch quietly to the dark bathroom, close and lock the door just as the footsteps reach my room. My thumb searches for the phone buttons in the dark as I push the small bathroom chair piled with red towels under the door handle. The door is heavy and strong, but the lock is flimsy. If the person on the outside wants in, I have about one minute.

I don't wait for niceties when I hear his voice. "Gil, it's me," I whisper. "I'm home. Someone is stalking me. Trying to get into the bathroom where I've locked myself in. Please hurry."

I hear the catch in his voice. "I'm close. Just a few minutes away; hold on."

I hear the footsteps explore the bedroom, pulling open the closet doors. Then the doorknob to the bath begins to jiggle, but the chair holds the lock in place. Jiggle again. Someone is trying to force it. It holds fast, thank God. Then silence. Maddening silence.

I think: The stalker is looking for something to use to beat the door in. Footsteps retreat, shuffling down the hall, moving toward the great room. I try to visualize the room. Sofa, chairs, table, fireplace. Oh God! The fireplace equipment! Plainly in view, heavy iron poker, shovel, tongs.

Silence. Then the footfalls begin again—heading toward me. I look at the bathroom window, painted shut probably fifty years ago. I look at the big old-fashioned claw foot bathtub with its heavy exposed pipes and faucets. Above it is a steel curtain rod attached to the ceiling in an oval loop on which the white canvas shower curtain slides around the tub.

As my eyes become accustomed to the dark, I see the outline of the showerhead with its steel coil. The coil can be stretched to about four feet as you move about, like a coiled phone cord. I reach for the bathtub faucet and turn it on hot. Then I detach the shower coil from its hook.

I can hear the pounding on the other side of the door, and know it's a matter of seconds before I am cornered. I flip on the shower head button at the top of the coil. The water is scalding. I hold it over the tub and then get a firm grip down low on the coil. I can swing the coil like a rope, a caveman weapon, a slingshot, David and Goliath. I don't know whether the stalker is armed or not, but I am. If the spurting, scalding water doesn't do it, a blow to the head with the heavy steel showerhead will be fatal.

"The police are on their way!" I scream as loud as I can over the roar of the water, but it sounds like a pitiful squeak. I keep shouting over and over: "The police are on their way, and I have a weapon. If you come in here I will kill you!" Even in my terror I cannot believe I hear myself say that. 'I will kill you! I will kill you!' And I mean it.

Now, why would this scene pretty much defy an outline? Because strict outlines in mysteries cut down on possibilities. The writer here didn't know how she was going to get Charm out of this situation. Seemingly the writer had painted herself into a corner. But ah! Visual and mental imagination got her out!

The writer imagined what was in that bathroom for Charm to protect herself with and the writer visualized an old-fashioned bathtub with a long, stretchable shower cord and a heavy shower head. Charm could turn on scalding water and sling the shower head at the stalker. Not the kind of thing a writer plans in an outline—until the material takes her there.

THE BIG DIG

That scene was a discovery in thinking through and fleshing out the action of a trapped character. In writing stories, you discover stuff all along the way—what the character is like, what the character would think and say, how he or she would say it. (The author first wrote this passage in the past tense: *I heard, I thought, I looked around the bathroom.* Then she reread it and began to think the stalker would be more menacing and Charm more frightened if the action were made immediate. So, she switched the verbs to present tense: *I hear, I think, I look . . .* Try reading it both ways and see whether you agree.)

When the character has a major decision to make, you, the writer, may discover something about yourself, as you figure out what the character would do. Remember in the old film, *High Noon,* where the male character, Will Kane, feels he has to uphold his code of honor and have a showdown with the bad guy? He does; he's a moral man. But in doing so he forces his Quaker wife, Amy, to violate her own moral principles—and shoot the bad guy herself, as it's the only way to save her husband's life. Who was the bigger hero here? You, the writer have to make Kane/Amy decisions in stories, and sometimes it will make you scratch your head and explore your conscience to do it.

Discovery is a big part of writing because every story, no matter how light or slight or funny, has some kind of moral core. It doesn't have to have an obvious "moral," as the aunt in Saki's story wanted to give her story, but a moral core is a sort of compass that sets us thinking: *What is right and what is wrong in this situation?* Sometimes in stories, the situation is so complex it's hard to tell; the solution develops *through our thinking and imagination*, right before our very eyes.

The Charm character in *The Bluebook Solution* excerpt had indicated earlier in the story that she did not believe in taking another life—not even a chicken's; hence, she's a vegetarian. But she had to make an instant decision, and says at the end of the passage, *"Even in my terror, I cannot believe I hear myself say that. 'I will kill you! I will* kill *you!'"* And even adds, *"And I mean it."*

REBOOT YOUR BRAIN

What will honorable people do to save their own lives? The lives of their family, kids, loved ones? What does "moral" mean? All good, and great, literature explores this question. *Good writers never tell you what to think; they tell you what's important to think about.*

What If

What if a widow with 4 starving children cannot leave them to earn money? What if small sums of money begin to disappear from a certain Mr. Big's bank account, and what if right after that, the widow's children get healthy and go back to school? What if an investigator suspects something? What if he explores the mystery?

What if, in the end (you supply the details), the investigator learns the widow did it? Does the investigator go to the cops? If he does, what do the cops do?

This is a simple SKELETON for a mystery short story, but it does have a moral core. Authors Charles Dickens and Victor Hugo wrote long, elaborate, complicated novels with a skeleton as simple as this.

What If

What if a character you create has just discovered that a crime has been committed and she knows who did it? What does she do? Roll over and play dead? Keep her mouth shut for fear of being ridiculed or implicated? Go to the cops? Try to solve it herself? None of the above?

Charm, in the excerpt you read, found herself in this situation. Charm, being Charm—the one the writer had created—*had* to go to the cops. They didn't believe her. So she headed out to solve the crime herself. Another character who wouldn't compromise her own principles. But then the non-killer Charm was all set to kill the stalker—a violation of her code of honor. When will people compromise their principles and when will they not? You, the writer, can set the questions in place for your characters and readers, then work through what you've set up to get the characters out of their situations—and hint to readers that this is an important idea to think about.

What If

What if somebody (you decide who) has committed a crime (you decide what)? What if the criminal has to face the victim and neither speaks a word, but both think through the crime and pertinent facts of their past lives? Get into the head of the criminal and the victim. What if, in this case, the criminal becomes more sympathetic to the reader than the victim? How do you, the writer, handle that? What happens in the end?

ZIP TIPS

Dicey Guys

Another good thing to remember: When you're writing stories with great *oppositions*, it's better not to have characters be obvious good guys (in old western stories, the white hat guys) and bad guys (the black hats). This makes your story simplistic (another genre in itself, the *melodrama*). And describing a character as "evil" isn't very helpful. It's such a broad term that it usually stops our thinking. And you, behind that zipper, want your reader thinking!

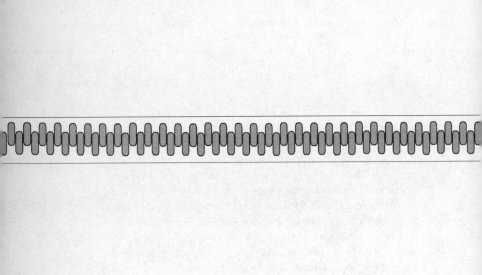

CHAPTER 7

Skeletons of a Feather:
Family Tales

FAMILY TALES CAN have complex and tragic endings or happy endings or humorous, twisted, or surprise endings. They can focus on one member of the family and see all the relationships and action from her or his point of view or the focus can be on a whole clan of people and how they interact. It can involve two families, maybe rivals, or those of different backgrounds, classes, or races. Basically, what the writer is telling is a story about several people related in some way and their behavior and interaction over a mutual incident or event. The death of Aunt Hattie and how each person in the family reacted to the event? The car crash that caused little brother Johnny to lose a leg? The will—and who gets what and why, and how does it affect each person? The list is endless. These short stories, along with some mysteries, often deal with psychological and social issues.

SPIN ZONE: GOODS THIS SHORT STORY NEEDS

Family stories obviously need families—two-person families, three, seven, whatever. Even extended families can be included, or two families, or several families in a neighborhood. Remember: the more people, the more the dynamics multiply. Family stories are good literary ways of dealing with race, gender, class, and other sociological issues. They always deal with emotional issues of some kind because home, as it has been said, is the place where, when you go there, "they" have to take you in. Maybe they don't want you for some reason; maybe you don't want them. Maybe they want you too much; maybe you need them too much. Maybe, actually, they won't let you in at all. Why? Even very short stories can explore some of this emotional dynamite.

Following is an excerpt from a family tale written by Ernest Hemingway and told by a limited omniscient narrator. Notice the family interaction and the attitude each member takes toward Harold Krebs, a son and brother, who has just returned from fighting in a war. The family lives in a small town in Oklahoma. Krebs has come back from war several months later than other American soldiers who were greeted as heroes. He doesn't think he's a hero; he's just a soldier who has done his duty. (The ellipses (dots) here indicate that a line or several have been omitted. This has been done for the purpose of space. But the omitted lines add no new information; they just repeat the same things Harold is thinking over and over).

Soldier's Home

. . . Nothing was changed in the town except that the young girls had grown up. But they lived in such a complicated world of already defined alliances and shifting feuds that Krebs did not feel the energy or the courage to break into it. He liked to look at them, though. There were so many good-looking young girls. Most of them had had their hair cut short. When he went away only little girls wore their hair like that or girls that were fast. They all wore sweaters and shirt waists with round Dutch collars. It was a pattern. He liked to look at them from the front porch as they walked on the other side of the street. He liked to watch them walking under the

shade of the trees. He liked the round Dutch collars above their sweaters. He liked their silk stockings and flat shoes. He liked their bobbed hair and the way they walked.

When he was in town their appeal to him was not very strong. He did not like them when he saw them in the Greek's ice cream parlor. He did not want them themselves really. They were too complicated. There was something else. Vaguely he wanted a girl but he did not want to have to work to get her. He would have liked to have a girl but he did not want to have to spend a long time getting her. He did not want to get into the intrigue and the politics. He did not want to have to do any courting. He did not want to tell any more lies. It wasn't worth it.

He did not want any consequences. He wanted to live without consequences. Besides he did not really need a girl. The army had taught him that. It was all right to pose as though you had to have a girl. Nearly everybody did that. But it wasn't true. You did not need a girl. That was the funny thing. First a fellow boasted how girls mean nothing to him, that he never thought of them, that they could not touch him. Then a fellow boasted that he could not get along without girls, that he had to have them all the time, that he could not go to sleep without them.

That was all a lie. It was all a lie both ways. You did not need a girl unless you thought about them. He learned that in the army . . . Now he would have liked a girl if she had come to him and not wanted to talk. But here at home it was all too complicated. He knew he could never get through it all again. It was not worth the *trouble. That was the thing about French girls and German girls. There was not all this talking . . . It was simple and you were friends. He thought about France . . . and Germany . . .*

He did not want to leave Germany. He did not want to come home. Still, he had come home. He sat on the front porch . . .

But he would not go through all the talking. He did not want one badly enough. He liked to look at them all, though. It was not worth it. Not now when things were getting good again.

He sat there on the porch reading a book on the war. It was a history and he was reading about all the engagements he had been in. It was the most interesting reading he had ever done . . . Now he was really learning about the war. He had been a good soldier. That made a difference.

One morning after he had been home about a month his mother came into his bedroom and sat on the bed. She smoothed her apron.

"I had a talk with your father last night, Harold," she said, "and he is willing for you take the car out in the evenings."

"Yeah?" said Krebs, who was not fully awake. "Take the car out? Yeah?"

"Yes. Your father has felt for some time that you should be able to take the car out in the evenings whenever you wished but we only talked it over last night."

"I'll bet you made him," Krebs said.

"No. It was your father's suggestion that we talk the matter over."

"Yeah. I'll bet you made him." Krebs sat up in bed.

"Will you come down to breakfast, Harold?" his mother said.

"As soon as I get my clothes on," Krebs said . . .

While he was eating breakfast his sister brought in the mail.

"Well, Hare," she said. *"You old sleepy-head. What do you ever get up for?"*

Krebs looked at her. He liked her. She was his best sister.

"Have you got the paper?" he asked.

She handed him The Kansas City Star *and he shucked off its brown wrapper and opened it to the sporting page. He folded* The Star *open and propped it against the water pitcher with his cereal dish to steady it, so he could read while he ate.*

"Harold," his mother stood in the kitchen doorway. *"Harold, please don't muss up the paper. Your father can't read his* Star *if it's been mussed."*

"I won't muss it," Krebs said.

His sister sat down at the table and watched him while he read.

"We're playing indoor over at the school this afternoon," she said. "I'm going to pitch."

"Good," said Krebs. "How's the old wing?"

"I can pitch better than lots of the boys. I tell them all you taught me. The other girls aren't much good."

"Yeah?" said Krebs.

"I tell them all you're my beau. Aren't you my beau, Hare?"

"You bet."

"Couldn't your brother really be your beau just because he's your brother?"

"I don't know."

"Sure you know. Couldn't you be my beau, Hare, if I was old enough and if you wanted to?"

"Sure. You're my girl now."

"Am I really your girl?"

"Sure."

"Do you love me?"

"Uh, huh."

"Will you love me always?"

"Sure."

"Will you come over and watch me play indoor?"

"Maybe."

"Aw, Hare, you don't love me. If you loved me, you'd want to come over and watch me play indoor." . . .

"You run along, Helen," [his mother] said. "I want to talk to Harold."

She put eggs and bacon down in front of him and brought in a jug of maple syrup for the buckwheat cakes. Then she sat down across the table from Krebs.

"I wish you'd put down the paper a minute, Harold," she said.

Krebs took down the paper and folded it.

"Have you decided what you are going to do yet, Harold?" his mother said, taking off her glasses.

"No," said Krebs.

"Don't you think it's about time?" His mother did not say this in a mean way. She seemed worried.

"I hadn't thought about it," Krebs said.

"God has some work for every one to do," his mother said. "There can be no idle hands in His Kingdom."

"I'm not in His Kingdom," Krebs said.

"We are all of us in His Kingdom."

Krebs felt embarrassed and resentful as always.

"I've worried about you so much, Harold," his mother went on. "I know the temptations you must have been exposed to. I know how weak men are. I know what your own dear grandfather, my own father, told us about the Civil War and I have prayed for you. I pray for you all day long, Harold."

Krebs looked at the bacon fat hardening in the plate.

"Your father is worried, too," his mother went on. "He thinks you have lost your ambition, that you haven't got a definite aim in life. Charley Simmons, who is just your age, has a good job and is going to be married. The boys are all settling down; they're all determined to get somewhere; you can see that boys like Charley Simmons are on their way to being really a credit to the community."

Krebs said nothing.

"Don't look that way, Harold," his mother said. "You know we love you and I want to tell you for your own good how matters stand. Your father does not want to hamper your freedom. He thinks you should be allowed to drive the car. If you want to take some of the nice girls out riding with you, we are only too pleased. We want you to enjoy yourself. But you are going to have to settle down to work, Harold. Your father doesn't care where you start in at. All work is honorable as he says. But you've got to make a start at something. He asked me to speak to you this morning and then you can stop in and see him at his office."

"Is that all?" Krebs said.

"Yes. Don't you love your mother, dear boy?"

"No," Krebs said.

His mother looked at him across the table. Her eyes were shiny. She started crying.

"I don't love anybody," Krebs said.

It wasn't any good. He couldn't tell her, he couldn't make her see it. It was silly to have said it. He had only hurt her. He went over and took hold of her arm. She was crying with her head in her hands.

"I didn't mean it," he said. "I was just angry at something. I didn't mean I didn't love you."

His mother went on crying. Krebs put his arm on her shoulder.

"Can't you believe me, mother? Please believe me."

"All right," his mother said chokily. She looked up at him. "I believe you, Harold."

Krebs kissed her hair. She put her face up to him.

"I'm your mother," she said. "I held you next to my heart when you were a tiny baby."

Krebs felt sick and vaguely nauseated.

"I know, Mummy," he said. "I'll try to be a good boy for you."

"Would you kneel and pray with me, Harold?" his mother asked.

They knelt down beside the dining room table and Krebs' mother prayed.

"Now you pray, Harold," she said.

"I can't."

"Do you want me to pray for you?"

"Yes."

So his mother prayed for him and then they stood up and Krebs kissed his mother and went out of the house. He had tried so to keep his life from being complicated. Still, none of it had touched him. He had felt sorry for his mother and she had made him lie. He would go to Kansas City and get a job and she would feel all right about it. There would be one more scene maybe before he got away. He would not go down to his father's office. He would miss that one. He wanted his life to go smoothly. It had just gotten going that way. Well, that was all over now, anyway. He would go over to the schoolyard and watch Helen play indoor baseball.

TONE ON TONE

Tone is a big part of this story. Notice how flat and repetitive the story is. A large part of this excerpt from the longer story takes place within the mind of Harold Krebs. We see that he thinks in circles and repeats himself.

In using this boring and repetitive tone, Hemingway is telling us something about Krebs' mental state. What might that be? Do you think Krebs could be depressed? He says he's angry at something. Could he have seen too much in the war? Maybe he is in denial—nothing is worth it, he keeps thinking, not even girls. Might Krebs be trying to get his head straight and thinks he is ridding himself of complications when—what happens? He's thrust back into the life and heart of his family.

Families are complicated. Krebs doesn't want that complication—all that talking, all that pretending to feel what he doesn't feel; he's seen too many lies. Even his sister wants him to be her "beau" and tell her he loves her. She talks too much and wants to implicate him, just like all the American girls he likes to look at but doesn't want. He wants an uncomplicated girl.

And his mother talks too much. She is trying to be sweet and gentle with him but she has no idea of her son's mental state. She lets him know that he's a

WORDS UNZIPPED

Interior Monologue: A monologue is one person or character speaking aloud. An interior monologue is one person or character thinking to himself or herself. These thoughts can be organized (as we speak) or rambling in a stream of consciousness (as our minds work, taking in many things at the same time).

subject of concern to her and his father and that they talk about him. Krebs doesn't want to be talked about; he wants to be left alone to sort out his life. His mother is not pushy, but she talks in sentimental clichés—sentimentality that Krebs, who has seen war, has no need for. The world has changed for him. He needs space, and time for healing, before he can re-enter this American world of his hometown.

And what else does his mother do? She treats him like a boy, tells him not to muss the paper, then lays a guilt trip on him. He was her tiny baby; he owes her something. Krebs allows her to reduce him to a child: *"I know, Mummy. I'll try to be a good boy for you."* An unseen, unheard Dad is even granting him permission to use the family car, as if he were a kid. Then Mummy makes him kneel and pray with her. He can't. He doesn't know what he's praying to or for. He's seen too many lies.

The *attitude* that Hemingway takes toward his subject (the protagonist, Krebs) creates the *tone* of the story, which, since this is such a static story, pervades the story itself. The only real action in the story is that which Krebs decides to take at the end of the excerpt—he *will* go away and get a job, he *will not* go see his father and endure another of these scenes, he *will* go to his sister's ballgame. This, by the way, is the end of the complete story.

At least he's made some kind of decision. The one-way conversation with his mother (notice he says very little) may not have produced the action she wants to see, but it at least got Krebs off the hook. And notice that the passive Krebs decides not only to take action but also to go see his sister's very active baseball game. So the reader sees some hope for Krebs' recovery from what we may perceive as his depression or post-traumatic stress disorder. Also note that when he enters the story, he is merely Krebs—because that's how he's been known in the military. We don't know his name is Harold until his mother uses it constantly, as one would with a child. His sister, with "Hare," splits the difference.

Though this story is told without emotion and with little description, lots of subtle things are going on. Read it again and visualize that you're seeing a good actor play Krebs on stage. Do you think it would be clear then what the point of the story is?

ZIP TIPS

Open Up to Me

Interior monologue (character talking to himself) is only one way to write a short story showing the interactions and emotions within families. There are many other ways. *Soldier's Home* relies a great deal on the author's tone—created by letting Krebs think. Even what little plot we find is created by the author's sustaining that tone.

Another writer, Gertrude Stein described the men who had served in World War I—they had to grow up fast "to find all Gods dead, all wars fought, all faiths in man shaken." Is this how a soldier might feel coming home?

Now here's another family story with a very different tone and voice. This one has a first-person narrator, the voice of the author of the book you're reading, and is based on a real experience. The setting is in Dallas, Texas, and the narrator and her husband, Greg, have just moved into a temporary rental house with their family, as they search for a new home to buy.

Foggy

The day we moved in, Greg had to go away on business. I began the chore of unpacking and putting the house together.

The kids ran in and out, sometimes helping, while I opened the packing boxes on the patio and hauled stuff inside to put away. I did this for about two days, bringing in take-out food, focused on trying to get the house in livable shape before the kids started to school.

Carrie came home from visiting neighbors and was reading in her room when she began to get sick. She complained, took an aspirin, and lay down to take a nap. Paul came home from across the street and soon began crying with a headache. He lay down on his bed. I continued to line the cupboards with insect-repellent paper and put dishes in the shelves. I noticed that the paper had a faintly odd odor to it, but it was hardly noticeable. Ben came in later, watched TV for awhile, then said he felt sick to his stomach and went to his bedroom.

After organizing the kitchen, I strode up and down the hall putting things away—towels, bed linens. I was in and out of the kids' rooms and noticed, vaguely, that all were taking naps, but their lassitude hardly registered. The door to the patio—where the packing boxes were stored—was open, so I moved in and out of the house for several hours.

It was late in the afternoon of the first or second day we'd been in the house (time grew a little foggy for awhile there) when I noticed that the kids were not up running around, looking in the fridge, flipping the TV channels. I noticed that, oddly, the TV wasn't even on at all.

A strange quietness surrounded me. I walked down the hall and looked in to see both boys asleep on their beds. Ben never slept! And Paul looked so pale. Ben's face was turned to the wall, and he was curled in a fetal position. Paul's mouth was open, and he was snoring a little, his eyes open a slit and rolled back in his head. These things registered, but not too clearly. I went on to Carrie's room, and she wasn't there. The house was dead still; my head felt funny. In the bathroom, I saw Carrie passed out on the floor in front of the toilet, as if she had gone there to throw up, but hadn't made it. I felt vaguely as if I were in a Twilight Zone movie.

I picked up Carrie and got her back to bed. Then I remember stopping again by the boys' door and looking in on them, half relieved that they were quiet for a change, but my head was in a fog. Back in the kitchen I decided irrationally that the odor of the insect-repellent paper in the cupboards was poisoned. I began to hallucinate. "This paper is killing my children," I muttered and began pulling it from the shelves. Plates and glasses crashed onto the floor as I pulled out the shelf paper. Nobody waked up. Suddenly, I could not see. It was as if a film had fallen over my eyes. Everything around me was dark and still. I fancied that I could smell the killer shelf paper enveloping me like some alien vapor. I was in the twilight zone. Something was sucking me into a vortex . . . I was going crazy . . . my kids were dying. I was on the floor on my hands and knees unable to see. The phone, I thought, the phone. Where is it?

I recalled a wall phone between the kitchen and breakfast room. I crawled to the area and began to climb my hands up the wall to feel for the phone. An outlet . . . a wire. Suddenly, there it was—the phone! Now, what was I to do with it? I knew no one in this city. My kids are dying, and who will I call?

I dialed the operator. "I am . . . I just moved to Dallas . . . I don't know anyone. I need a doctor. I need help. My children and I have been poisoned. Please help me."

The operator asked the address. "I don't know a doctor, but there's a drugstore near you. Just a moment."

In a few seconds a druggist came on the line. "I don't know," I said. "Just moved here . . . know no one . . . poisoned . . . kids are dying . . . I'm blind . . . please help."

"Where exactly are you?" he asked. I told him as well as I could. "I'll get someone there," he said. "Hang on."

I don't know how long it was, but a doctor came to the door. I heard the knock but couldn't get there. I opened my eyes but couldn't see. I began to crawl to the door.

I heard it open; footfalls came toward me. "What's wrong here?" a male voice said. I remember feeling the cool hard wood of the floor against my cheek. Someone is here; now I can die.

The doctor shook me. "Who else is here?" he asked loudly. "Tell me!" I raised my head. "My kids . . . bedroom . . . boys . . . Carrie . . . back room. . ." I fell to the floor again.

He shook us all awake and got us into the living room. "I think you must have food poisoning," he said. "No, it has nothing to do with 'poison' shelf paper." He gave us pills and water. He slapped our cheeks. "Now, don't get dehydrated. I'll send a home nurse here every six hours to make sure you get enough liquid. She'll be here right away and stay with you for awhile. You'll be okay." He made a phone call and left.

The nurse came and fed us liquids. Water, cokes, ice tea. "Just keep drinking. Doctor says you'll feel better." She came that night and had us up walking around the yard. The fresh air seemed to revive us; my sight returned briefly. She came again the next morning and made us drink through straws. When we fell over, she propped us up and gave us liquid. She walked us around outside. Then she left.

The second or third day—I lost count of time—I called my mother, who lived about two hours away. No answer. I called Greg's mother, who lived about the same distance. "We are very sick, and I need help," I said. "I don't know what's wrong, but Carrie is passed out in the hall again, and the boys haven't moved for hours. I don't know how many. I can't see and can hardly hold my head up."

I crawled over to the sliding door of the patio and inched them open. Then I pulled myself outside, hunching along like some demented Caliban, and put my face against the stones of the patio floor. I felt better almost immediately. This house is killing us. I have to get my kids out. I hauled myself back through the door and down the hall. I took Paul, the smallest, first, dragging him off the bed. He was barely conscious. Creeping along together we got outside to the patio. I went back for Ben, and then Carrie— dragging them down the hall and out the door, as they half-cried, half-whimpered. The image of a mother dog rescuing her puppies flashed into my head.

On the patio outside, the temperature was in the high 90s. Inside, the house was air-conditioned, but we felt better in the heat. Within a few hours Greg's mother and father, Trudy and Will, bustled in with liquids and food. I was not hungry at all, but my limbs felt like stone. Trudy put me to bed. I remember lying there imagining that I was the Sphinx with massive, heavy, concrete limbs. Then I remember nothing at all . . . until . . .

Something roused me in the night. I will never know what. Mere survival instinct, maybe. I pulled myself out of bed and again half-crawled down the hall. My eyesight would come and go in a blur. I saw Trudy, in a pink nightgown, passed out near a bedroom door. Will was asleep on the floor. In the living room, Greg, still in shirt and tie, was in a deep sleep on the sofa. I didn't know when Greg had arrived, but I couldn't rouse him.

By this time I was sure the house was cursed, so I pulled my body to the phone and called the operator to get an ambulance. "Everyone is dead or dying!" I screamed. "Hurry! Please hurry!"

Within a few minutes I heard the siren. Then knocks at the door. Then people walking, men talking. I lifted my head and saw my kids in a blur, one by one, carried out the door. Carrie's little sun-tanned arms dangling down; Ben's strong, active body, now limp; Paul's eyes turned in on themselves. These images registered, but I could not move. Then I saw a white-coated man carrying Trudy in her pink nightgown out the door, followed by a man holding Will in his striped pajamas. Greg in his suit pants and dress shirt, tie awry, went by in the arms of a burly man. Then they came to the kitchen for me.

We all lay on stretchers in the front yard. Sirens blared, red lights flashed. It seemed the world was filled with raucous, confusing alarms. Somehow I found Ben's hand. Then we were all lifted into the ambulance. I remember thinking with a fuzzy horror: Are we going to be on TV like this? New School Chief Arrives in City, Dies in Front Yard . . .

At the hospital, I recall seeing Greg in a wheelchair. He was being rolled to a desk to register us, but he couldn't speak. His head had fallen onto his chest. I think I was walking by this time, but don't remember exactly. I do recall telling the desk person that my family was dying, to take care of them, and I'd settle it up with her about the insurance later. I think I was screaming. They put us all together in one big room. Doctors and nurses hovered over us. They gave us liquids, and we lay there for how long I don't know. Little by little we began to feel better. I

MORE TO UNZIP
WHAT IF

What if a man sits in his car just across a small lake from his brand new house, which he is proud of being able to buy? In a fantasy, he sits and imagines the kind of family who might live there. He comes home, and his wife, or kids, or all of them do something that remind him precisely of the vision he just had. What was he thinking? What did they do?

What if an American Indian family is having dinner on what most of us call Thanksgiving Day? What's their conversation? What are their thoughts?

What if a young man whose parents have been killed in an accident looks through old family albums as he sorts out their belongings? Though there are many photos of the other children in the family, there are none of him. Why?

What if the children of an older couple are in their old home together for a Christmas visit? The parents entice their children to go to the church of their childhood with them. The children are grown, maybe married into another faith, maybe married into another race. What is each child thinking? Are these thoughts sweet and nostalgic? Bitter? Rebellious?

Rattle those bones, guys. It's fun.

remember opening my eyes and seeing little skinny, darling Paul sitting up on the white cot, yawning and looking around as if asking: How did I get here? We awoke gradually, first one and then another. We mumbled at each other a little, wondering what was wrong, why we were here. The kids were surprised to see their grandparents, and especially laid out on cots as they were. They were the last to rouse themselves.

At some point a doctor came in and looked in our eyes with a light. "You're very lucky," he said, patting me on the cheek. "We sent a city inspection team to your house. You've been getting lethal dosages of carbon monoxide from a water heater in the hall. The

ZIP TIPS

Don't Try This at Home

Here's a dramatic incident in the life of a family, told with almost reportorial calmness, after-the-fact, by the narrator/protagonist. What force was the antagonist? The mystery is that we don't know until the very end, when we learn of the carbon monoxide. The suspense builds until then. The readers are puzzled right along with the narrator. These things are fun to manipulate and experiment with. Try it, you'll like it. Writing is playing.

heater was not plumbed properly, and all the toxic air was being filtered into the house. I'm surprised you're still alive."

So that's why we felt better lying on the hot patio, I thought later. Each time I went outside, I began to get my sight back; then lost it again inside. At this time, I had never heard of carbon monoxide poisoning, and the city inspector confirmed that such incidents were very rare. Well, I thought, I had sprung from my mother's womb eager for rare adventures, but this was a little more than I'd bargained for.

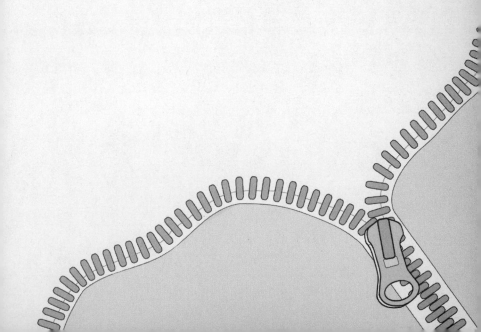

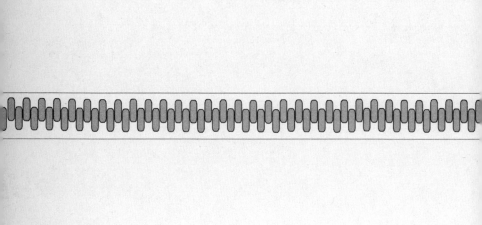

CHAPTER 8

Kissing Skeletons:
Romance Stories

ROMANCES, WHETHER SIMPLE or complicated, have an obstacle to overcome before the couple comes together or drifts apart. What is it? That's for you, the writer, to imagine. In vivid, specific words, you must make the story visual and believable and you must provide momentum. In *Romeo and Juliet,* the obstacle is feuding families. Other obstacles might be another man or woman, an existing wife or husband, a wicked stepsister, a bad-seed child, or a condition—one is going off to war, one has a fatal disease or mental illness. One is of the conventionally wrong sex or race or religion. Any of these can get in the way of whatever the two want to do together. The writer devises a way of allowing them to overcome the obstacles or accept the fate that keeps them apart. Love has to be tested or it isn't love.

GREAT HAIR AND TEETH: GOODS THIS SHORT STORY NEEDS

Characters are frustrated because someone or something gets in their way (plot, though, is secondary). Remember the magic number three? It takes at least 3 people to make these stories work. The "needs" are love, love denied, lost, recaptured, and even love lost again. We must like the characters and see them as unique people, able to make the story either comic or tragic. Sometimes you can write a story where the characters DO NOT have great hair and teeth, and that's part of the point of the story. Maybe they're two physically ugly characters, and they know it, but something the writer tells us about their psychology makes us love them and want them to win. WE CAN'T ALL BE HOTTIES! It's an open-ended genre, so your unzipped imagination can do anything with it.

YOU WANT DRAMA QUEENS OR DRAMA?

In writing romance (and family stories)—those that tend to tip the seesaw to favor Character—you have something else to decide, too. For example, what *kind* of romance do you want to write? Do you want to write the *Love Story* kind of romance—a fairly predictable tear-jerker? Or the Hollywood happy-ending kind of romance, such as *When Harry Met Sally* meets *Sleepless in Seattle,* and *Pretty Woman* meets *Notting Hill?* Robert Altman's film, *The Player,* is a vicious satire of these sentimental, happy, escape-life kinds of films. If you want to write this kind of romance, have at it. There is a place for it, and, as the writer, you're the boss. Much current "chick lit" and many fairy tales are this kind of romance.

However, if you want to write a more sophisticated kind of romance or love story, think of *Casablanca.* This film, often called the best film in history, does not have a happy, love story ending, and it is full of *sentiment* (gritty, complicated, real human emotion), rather than *sentimentality* (clichés, predictable, manu-factured emotion).

Neither choice is right or wrong, but your choice depends on the purpose of your story and the audience to which you want to appeal.

A novelist and short story writer, Lee Smith, is definitely a writer of gritty, complicated, real human emotions who appeals to serious readers. In the excerpt following, she is having fun writing a *satire* (a spoof, a take-off, an over-the-top mockery) of clichéd, predictable romance.

She begins by telling us in the Preface that a friend of hers decided to write a romance, so she wrote to Silhouette Romances for guidelines (a real publishing company; google it for more information). Her friend, Smith writes, decided not to write the story after reading the guidelines, so Smith herself did. Here are the guidelines Smith's friend received:

Our Heroine is, preferably, an orphan. She is alone in the world. (Note: A brother is, in some cases, permissible, but only if he is retarded or has not found his way in life.) Our Heroine appears frail, but looks terrific when she gets dressed up. She is, of course, a virgin. She arrives alone in the lush, romantic Setting, where she encounters our Hero, who is preferably dark, brooding, and mysterious (although we have had some luck recently with stern Nordic sorts and hunky redheads). The initial encounter is tempestuous. Sparks fly, yet there is of course a mad, underlying attraction. The Other Woman will be beautiful, desirable, and wealthy. She is, of course, a bitch. The Other Man will be nice, boring, well-meaning, intent upon saving our Heroine from the clutches of our Hero and the dangerous contingencies of the Plot. (Note: No other main characters will be permitted in this novel, especially children. Any necessary others, such as a faithful housekeeper, should remain as stereotypical as possible, so as not to detract from the romance.) The Plot will ensue, with the ten chapters growing increasingly shorter as tension mounts. At the climax, our Hero and Heroine realize that they are made for each other after all. The novel ends with their passionate embrace. (Note: at no time during this novel will they or anyone else ever actually do it, nor will any specific body parts be mentioned.)

Cheesy

How would you describe a story written according to these instructions? You'd be right, if you said "formula fiction." That's exactly what this is, a formula for writing from which the writer is not allowed to deviate. It can lead to some really deadly, or laughable, writing, as Smith goes on to illustrate in her story, written to this formula. The following sample from the first chapter is enough to prove her point.

Desire on Domino Island

. . . Jennifer surveys the lush scene before her with no small trepidation, and a hint of dismay creeps into her normally dulcet tone as she exclaims, "Captain! Oh, Captain! Why are you docking here in the middle of nowhere? . . ."

But the captain won't say a thing! A native Georgian with an unfortunately cleft palate, he shoots a dark glance from beneath his surly brow at the clearly frightened young woman and mumbles something indistinguishable into his dark facial hair. He throws her bags on the beach. He heaves his bulk around.

Jennifer drums her small fingers rat-a-tat-tat on the hull of the shiny craft. Is it all a huge mistake, her coming here? But what else could she have done, considering the terrible fire that swept the home of her guardians (since their parents' mysterious death some twenty years ago, Jennifer and her retarded brother, Lewis, have been most carefully raised), killing both Aunt Lucia and Uncle Norm and destroying the entire perfect loveliness of their antebellum mansion, leaving Jennifer with only her small inheritance, her paltry background in microbiology, and the hunting lodge somewhere deep within the vastness of this fabled island . . .

Diva Lit

Well, it goes downhill from here. But you can see what Smith is doing—playing, tongue-in-cheek, with the strict formula and creating pretty silly characters. Also notice that she is *telling* not *showing*, something that she knows a good writer does not do. And notice, too, that along with all the clichéd characters—orphaned protagonist, dark, surly man, retarded brother, antebellum mansion, small inheritance, hunting lodge, and fabled island—the writer throws in a *"paltry background in microbiology"*—the only fresh, un-clichéd detail in all this nonsense, the only surprise, and the only bit of humor. You can be assured that this *"paltry background,"* or Jennifer's use of its sophisticated content, will never be heard from again in this story. Clearly, Smith is deliberately violating the formula by making Jennifer a tad interesting. (*No! No! Not acceptable, according to the guidelines!*) This style of writing often uses many exclamation points!!! Poor writers do this because they don't have the skills to build the energy and excitement into their sentences and images. Overuse of exclamation points signals manufactured emotion.

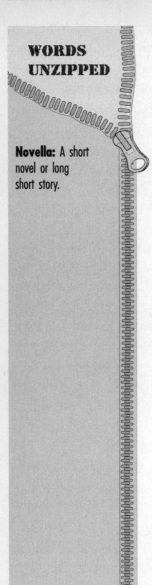

WORDS UNZIPPED

Novella: A short novel or long short story.

DIRTY, DIRTY!

Next, read an excerpt from another serious writer, like Smith, who long ago played with the formula in a different way. D.H. Lawrence, in 1930, wrote a novella, *The Virgin and the Gipsy,* using basically the same Silhouette formula (before it was ever formulated). His Heroine is fair, lovely, frail ("but looks terrific when she gets dressed up"). She is "of course a virgin," as the title indicates. The Hero (the *Gipsy*—Lawrence spells it this way—is so mysterious that he does not even have a name until the last two words in the story) is "dark and brooding." The initial encounter is tempestuous. (Meeting the irresistible Hero, the Heroine has a woman gypsy tell her fortune, as she leaves her "nice, boring, well-meaning" boyfriend, Leo, and other clichéd buddies in the car waiting). The "Other Woman" is the gipsy's wife, who hardly appears, except as possibly the fortune teller, so we don't know whether she's a "bitch." And the "faithful housekeeper" is a dark, nasty, "stereotypical" Aunt Cissie. The setting is a small village in the "lush" English countryside, where

Yvette (our Heroine) is one of two daughters of the local, very uptight Vicar (a minister). Lucille is the older sister, but she stays in the background and certainly does not upstage our Heroine, Yvette.

The Virgin and the Gipsy

. . . Leo sounded the horn sharply. The man in the cart looked round, but the woman on foot only trudged steadily, rapidly forward, without turning her head.

Yvette's heart gave a jump. The man on the cart was a gipsy, one of the black, loose-bodied, handsome sort. He remained seated on his cart . . . and his pose was loose, his gaze insolent in its indifference. He had a thin black moustache under his thin, straight nose, and big silk handkerchief of red and yellow tied round his neck . . . Leo made the horn scream . . . The gipsy turned round at the din, laughing in his dark face under his dark-green cap, and said something which they did not hear, showing white teeth under the line of black moustache, and making a gesture with his dark, loose hand.

"Get out o' the way then!" yelled Leo.

For answer, the man delicately pulled the horse to a standstill, as it curved to the side of the road. It was a good roan horse, and a good, natty, dark-green cart.

Leo, in a rage, had to jam on the brake and pull up, too.

"Don't the pretty young ladies want to hear their fortunes?" said the gipsy on the cart, laughing except for his dark, watchful eyes, which went from face to face, and lingered on Yvette's young, tender face.

She meets his dark eyes for a second, their level search, their insolence, their complete indifference to her breast. (Oops! A body part! Lawrence violated the formula!) She thought: "He is stronger than I am! He doesn't care!"

"Oh yes! Let's!" cried Lucille at once.

"Oh yes!" chorused the girls . . .

This is enough to give you, but the story goes on. The fortune teller presumably tells Yvette that a dark, handsome, brooding stranger will enter her life, and she is to *"listen for the water."* If this romance makes you curious, and captures your fancy, read the entire story. The thing is, what Lawrence does here is take the formula story and, rather than stop with that, he makes the characters *symbolic;* that is, they are actual, realistic people, but they *represent* two different ways of thinking and being. One is conventional, establishment, uptight, and righteous, like Yvette's family and friends. One is wild, exploratory, "loose" (as you, several times, see the gipsy to be), and, as Lawrence would have us believe, more in touch with nature and his feelings. The story goes on, in its formulaic way, and the crush Yvette develops for the gipsy is told mainly in her mind.

The formula is followed: they are pristine in their attraction for one another; they do not "do it." And in the end, the gipsy rescues

Yvette from an awful flood (*listen for the water*) and finally has to dry her down with a towel (oh, how intimate!) and warm her body with his to keep them both from freezing.

But the story is lifted from the formula by the author's ability to make his characters not only people, but also to take them to another level in making them represent ideas. It's a story about cultural constraints and what they do to us, as well as a romance story and what romance does to us.

NAME DROPPING

You may want to write a formula romance. Or you may want to take the story to a deeper level. Or you may want to do something altogether different, as in the following story excerpt by Anton Chekhov.

The Kiss

Six batteries of an Artillery Brigade plan to spend night in a village . . . before setting up camp. The officers receive an invitation from a local general, General von Rabbek, to come to tea at his estate. Though they are tired, they cannot refuse. They start on their way. Soon they see . . . the lighted windows of von Rabbek's house. One officer says,

"Yes, there are women here. My instinct tells me . . ."

Von Rabbek . . . pressed their hands . . . In a big dining room, at a big table, sat ten men and women, drinking tea . . . The visitors, some with serious, even severe faces, some smiling constrainedly, all with a feeling of awkwardness, bowed, and took their seats at the table. Most awkward of all felt Staff-Captain Riabovich, a short, round-shouldered, spectacled officer, whiskered like a lynx. While his brother officers look serious . . . his face, his lynx whiskers and his spectacles seemed to explain: "I am the most timid, modest, undistinguished officer in the whole brigade."

. . . After tea the guests repaired to the drawing room . . . packed with young women and girls . . . Music began . . . They began to dance. Young von Rabbek valsed twice round the room with a very thin girl; and Lobuitko, slipping on the parqueted floor, went up to the girl in lilac, and was granted a dance. But Riabovitch stood near the wallflowers, and looked silently on. Amazed at the daring of men who in sight of a crowd could take unknown women by the waist, he tried in vain to picture himself doing the same . . . With years he had grown reconciled to his own insignificance, and now looking at the dancers and loud talkers, he felt no envy, but only mournful emotions.

At the first quadrille von Rabbek junior approached and invited two nondancing officers to a game of billiards . . . Riabovitch, who stood idle . . . followed . . . Riabovitch, whose only game was cards, stood near the table and looked indifferently on . . . Before the game was over he was thoroughly bored, and, impressed by a sense of his superfluity, resolved to return to the drawing room and turned away.

It was on the way back that his adventure took place. Before he had gone far he saw that he had missed his way . . . Retracing his steps, he turned to the left, and found himself in an almost dark room which he had not seen before; . . . Riabovitch paused in irresolution. For a moment all was still. Then came the sound of hasty footsteps; then, without any warning of what was to come, a dress rustled, a woman's breathless voice whispered, "At last!" and two soft, scented, unmistakably womanly arms met round his neck, a warm cheek impinged on his, and he received a sound kiss. But hardly had the kiss echoed through the silence when the unknown shrieked loudly, and fled away—as it seemed to Riabovitch—in disgust. Riabovitch himself nearly screamed and rushed headlong towards the bright beam in the door chink.

As he entered the drawing room his heart beat violently, and his hands trembled so perceptibly that he clasped them behind his back. His first emotion was shame, as if everyone in the room already knew that he had just been embraced and kissed . . . His neck, fresh from the embrace of two soft, scented arms, seemed anointed with oil; near his left mustache, where the kiss had fallen, trembled a slight, delightful chill, as from peppermint drops; and from head to foot he was soaked in new and extraordinary sensations, which continued to grow and grow.

He felt that he must dance, talk, run into the garden, laugh unrestrainedly. He forgot altogether that he was round-shouldered, undistinguished, lynx whiskered, that he had an "indefinite exterior"—a description from the lips of a woman he had happened to overhear. As Madame von Rabbek passed him he smiled so broadly and graciously that she came up and looked at him questioningly.

"What a charming house you have!" He said, straightening his spectacles. And Madame von Rabbek smiled back . . .

At supper Riabovitch . . . devoted all his powers to the unraveling of his mysterious, romantic adventure. What was the explanation? It was plain that one of the girls, he reasoned, had arranged a meeting in the dark room, and after waiting some time in vain had, in her nervous tension, mistaken Riabovitch for her hero . . . "But which of them was it?" he asked, searching the women's faces. She certainly was young . . . Secondly, she was not a servant. That was proved unmistakably by the rustle of her dress, the scent, the voice.

When at first he looked at the girl in lilac, she pleased him; she had pretty shoulders and arms, a clever face, a charming voice. Riabovitch piously prayed that it was she . . . Riabovitch turned his eyes on the blonde in black. The blonde was younger, simpler, sincerer; she had charming kiss-curls, and drank from her tumbler with inexpressible grace. Riabovitch hoped it was she . . . "It is a hopeless puzzle," he reflected. "If you take the arms and shoulders of the lilac girl, add the blonde's curls, and the eyes of the girl on Lobuitko's left, then . . ."

Supper over, the visitors, sated and tipsy, bade their entertainers good-by . . . On reaching home, he undressed without delay and lay upon his bed . . . "Where is she now?" muttered Riabovitch, looking at the soot-blacked ceiling.

His neck still seemed anointed with oil, near his mouth still trembled the speck of peppermint chill. Through his brain twinkled successively the shoulders and arms of the lilac girl, the kiss-curls and honest eyes of the girl in black, the waists, dresses, brooches. But though he tried his best to fix these vagrant images, they glimmered, winked, and dissolved; and as they faded finally into the vast black curtain which hangs before the closed eyes of all men, he began to hear hurried footsteps, the rustle of petticoats, the sound of a kiss . . .

Riabovitch pulled the bedclothes up to his chin, curled himself into a roll, and strained his imagination to join the twinkling images into one coherent whole. But the vision fled him. He soon fell asleep, and his last impression was that he had been caressed and gladdened, that into his life had crept something strange, and indeed ridiculous, but uncommonly good and radiant . . .

When he awoke the feeling of anointment and peppermint chill was gone. But joy, as on the night before, filled every vein. He looked entranced at the windowpanes gilded by the rising sun, and listened to the noises outside. Someone spoke loudly under the very window. It was Lebedietsky, commander of his battery, who had overtaken the brigade. He was talking to the sergeant-major . . .

Fifteen minutes later the brigade resumed its march. As he passed von Rabbek's barns, Riabovitch turned his head and looked at the house . . . Evidently, all still slept . . . Among them slept she—she who had kissed him but a few hours before. He tried to visualize her asleep. He projected the bedroom window opened wide with green branches peering in, the freshness of the morning air, the smell of poplars, lilacs, and roses, the bed, a chair, the dress which rustled last night, a pair of tiny slippers, a ticking watch on the table . . . But the features, the kind, sleepy smile—all, in short that was essential and characteristic—fled his imagination as quicksilver flees the hand . . .

Towards evening the brigade ended its march . . . Riabovitch, whose head was dizzy from uninterrupted daydreams, ate in silence. When he had drunk three glasses, he felt tipsy and weak; and an overmastering impulse forced him to relate his adventure to his comrades.

"A most extraordinary thing happened to me at von Rabbeck's," he began ". . . I was on my way, you understand, from the billiard room . . ."

And he attempted to give a very detailed history of the kiss. But in a minute he had told the whole story. In that minute he had exhausted every detail; and it seemed to him terrible that the story required such a short time. It ought, he felt, to have lasted all the night. As he finished, Lobuitko, who as a liar himself believed in no one, laughed incredulously. Merzliakoff frowned, and . . . said indifferently—"God knows who it was! She threw herself on your neck, you say, and didn't cry out! Some lunatic, I expect!"

"It must have been a lunatic," agreed Riabovitch.

"I, too, have had adventures of that kind," began Lobuitko, making a frightened face. "I was on my way to Kovno . . ." The

story annoyed Riabovitch. He rose from the box, lay on his bed, and swore that he would never again take anyone into his confidence . . .

On the thirty-first of August he left camp . . . He longed passionately for . . . the dark room; and that internal voice which so often cheats the lovelorn whispered an assurance that he should see her again. But doubt tortured him. How should he meet her? What must he say? Would she have forgotten the kiss? If it came to the worse—he consoled himself—if he never saw her again, he might walk once more through the dark room, and remember . . .

"When von Rabbek hears from the peasants that we are back he will send for us," thought Riabovitch . . . impelled by restlessness, he went into the street and walked . . . When he reached von Rabbek's garden Riabovitch peered through the wicket gate. Silence and darkness reined . . . Riabovitch listened greedily and gazed intently. For a quarter of an hour he loitered, then hearing no sound and seeing no light, he walked wearily towards home . . .

"How stupid! How stupid!" thought Riabovitch . . . "How stupid everything is!"

Now that hope was dead, the history of the kiss, his impatience, his ardor, his vague aspirations and disillusion appeared in a clear light. It no longer seemed strange that the general's horseman had not come, and that he would never again see her who had kissed him by accident, instead of another . . . And the whole world—life itself—seemed to Riabovitch an inscrutable, aimless mystification

. . . He recalled how Fate in the shape of an unknown woman had once caressed him; he recalled his summer fantasies and images—and his whole life seemed to him unnaturally thin and colorless and wretched . . .

When he reached the cabin his comrades had disappeared. His servant informed him that all had set out to visit "General Fonrabbkin," who had sent a horseman to bring them. For a moment Riabovitch's heart thrilled with joy. But that joy extinguished. He cast himself upon his bed and wroth with his evil fate, as if he wished to spite it, ignored the invitation.

Well, there's a story that resides a good deal inside the mind of the protagonist, told by a limited omniscient narrator. What is the obstacle to Riabovitch's love? How does he deal with it? Why do you suppose he grew so obsessed with the girl and the kiss? Does it seem that Riabovitch's reaction was disproportionate to its cause? If he had been characterized as a different kind of person, would he have reacted in the same way? Do you think that if he had discovered who kissed him, he would have reacted differently? Did he lie to his comrades by not relating that the kissing girl shrieked loudly afterward? What would lead us to believe that he lied? Why did he choose to ignore the next invitation? Would it be that Riabovitch was afraid to test this "love"?

In this story, the author attempts to trace a psychological reaction, grounded in a character. Did you notice how he changed after the kiss? Then did you notice how he changed again, after deciding the whole thing was hopeless? Do you think this is an apt psychological description of someone with a strong crush on another person?

WHAT IF

What if a boy and girl meet and you want to write a formulaic story? In what lush place do you set them? In what pristine ways do they conduct themselves? What will be the obstacle? Will they overcome it or accept their fate? What drama, action, conflict can you create from this?

ZIP·TIPS

Shotgunning

One of the best ways to get yourself into writing is this little trick: Shoot your ideas directly out of your head, as if you were shooting a gun. Just pull the trigger and watch your fingers as they type words, ideas, sentences. Don't monitor your thoughts. Just start thinking of the story you want to write and let your mind roll. Don't correct anything. Don't spell-check. Don't stop and grope for a word. Keep those words coming until you've filled a couple of pages. Then take a break and come back and read it. You'll see that the act of writing actually helps you think. You'll clean it all up later—the shells, the scattered shots, the stuff that's too random. But you'll have unzipped your mind.

What if a boy meets a girl and she reminds him of his beloved, but dead, mother, domestic and nurturing? The girl meets the boy and sees in him the strong-minded, independent, worldly, intellectual man she had wanted her father to be. The two decide to marry. Neither turns out to be as the other saw him or her. What happens?

What if a boy from the lower classes meets a girl from the middle classes? They fall in love. Why? Why were they attracted to one another? They marry and have children. He wants the kids to remember where their daddy came from and not rise above their class. She wants the kids to know what is out there for them with her connections. What happens?

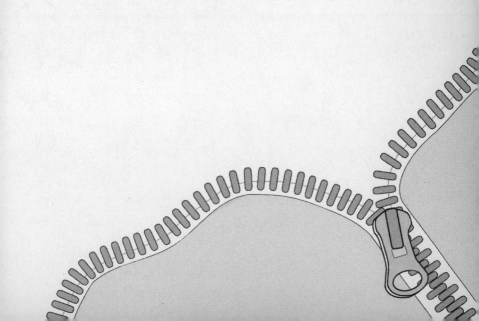

CHAPTER 9

Creaky Skeletons:
Horror Stories

THESE STORIES NEED some really frightening, grotesque, haunting, horrifying, grisly, even disgusting, yucky element to scare readers out of their shoes. Stephen King, the novelist, knows how to do this as well as anyone. Thomas Harris, in *Silence of the Lambs*, does it in a disgusting way—the bad guy cannibalizes his victims. Horror stories can begin just about anywhere in the tale and end anywhere—as long as they have bloody bones, or people who eat people, or talking corpses, or werewolves, or fanged bats, or even mad murderers. Anyone who has ever told a spooky story to terrify a little brother or sister at night can write a short horror story.

Here's an excerpt from a horror short story by Edgar Allan Poe, the master of stories that creep us out. It's told in the first person by a man who has murdered another much older man the narrator is supposed to be taking care of.

The Tell-Tale Heart

. . . It is impossible to say how first the idea entered my brain; but once conceived, it haunted me day and night . . . I think it was his eye! Yes, it was this! One of his eyes resembled that of a vulture—a pale blue eye, with a film over it. Whenever it fell upon me, my blood ran cold, and so by degrees—very gradually—I made up my mind to take the life of the old man, and thus rid myself of the eye forever . . .

You fancy me mad. Madmen know nothing. But you should have seen me. You should have seen how wisely I proceeded—with what caution—with what foresight . . . I was never kinder to the old man than during the whole week before I killed him. And every night, about midnight, I turned the latch of his door and opened it—oh, so gently! And then, when I had made an opening sufficient for my head, I put in a dark lantern, all closed, closed, so that no light shone out, and then I thrust in my head. Oh, you would have laughed to see how cunningly I thrust it in! I moved it slowly—very, very slowly, so that I might not disturb the old man's sleep. It took me an hour to place my whole head within the opening so far that I could see him as he lay upon his bed. Ha!—would a madman have been so wise as this? And then, when my head was well in the room, I undid the lantern cautiously—oh, so cautiously—cautiously (for the hinges creaked)—and I undid it just so much that a single thin ray fell upon the vulture eye. And this I did for seven long nights—every night just at midnight—but I found the eye always closed; and so it was impossible to do the work, for it was not the old man who vexed me, but his Evil Eye . . .

Upon the eighth night I was more than usually cautious in opening the door. A watch's minute hand moves more quickly than did mine. Never before that night had I felt the extent of my own powers . . . my feelings of triumph. To think that there I was, opening the door, little by little, and he not even to dream of my secret deeds or thoughts. I fairly chuckled at the idea; and perhaps he heard me, for he moved on the bed suddenly, as if startled . . . His room was black as pitch with the thick darkness . . .

I had my head in, and was about to open the lantern, when my thumb slipped upon the tin fastening, and the old man sprang up in the bed, crying out "Who's there?" . . .

I did not hear him lie down. He was still sitting up in the bed listening just as I have done, night after night . . .

Presently I heard a slight groan, and I knew it was the groan of mortal terror. It was not a groan of pain or of grief—oh, no! It was the low, stifled sound that arises from the bottom of the soul when overcharged with awe. I knew the sound well. Many a night just at midnight, when all the world slept, it was welled up from my own bosom, deepening with its dreadful echo, the terrors that distracted me. I knew what the old man felt and pitied him, although I chuckled at heart. I knew that he had been lying awake ever since the first slight noise, when he had turned in the bed. His fears had been ever since growing upon him. He had been trying to fancy wind in the chimney—it is only a mouse crossing the floor . . . Yes, he had been trying to comfort himself with these suppositions; but he had found all in vain. All in vain; because Death, in approaching him, had stalked with his black shadow before him and enveloped the victim. And it was the mournful influence of the unperceived shadow that caused him to feel—although he neither saw nor heard—to feel the presence of my head within the room.

When I had waited a long time . . . I resolved to open a little—a very, very little crevice in the lantern. So I opened it—you cannot imagine how stealthily, stealthily, until, at length, a single dim ray, like the thread of a spider, shot from out the crevice and full upon the vulture eye.

It was open—wide, wide open—and I grew furious as I gazed upon it. I saw it with perfect distinctness—all a dull blue, with a hideous veil over it that chilled the very marrow in my bones; but I could see nothing else of the old man's face or person; for I had directed the ray as if by instinct, precisely upon the damned spot.

And now, have I not told you that what you mistake for madness is but over acuteness of the senses? Now, I say, there came to my ears a low, dull, quick sound, such as a watch makes, when enveloped in cotton. I knew that sound well too. It was the beating of the old man's heart. It increased my fury, as the beating of a drum stimulates the soldier into courage.

. . . The hellish tattoo of the heart increased. It grew quicker and quicker, and louder and louder every instant. The old man's terror must have been extreme! It grew louder, I say, louder every moment!—do you mark me well? I have told you that I am nervous; so I am. And now at the dead hour of the night, amid the dreadful silence of that old house, so strange a noise as this excited me to uncontrollable terror. Yet, for some minutes longer I refrained and stood still. But the beating grew louder, louder! I thought the heart must burst. And now a new anxiety seized me—the sound would be heard by a neighbour! The old man's hour had come! With a loud yell, I threw open the lantern and leaped into the room. He shrieked once—once only. In an instant I dragged him to the floor and pulled the heavy bed over him. I then smiled gaily, to defend the deed so far done. But, for many minutes the heart beat on with a muffled sound. This, however, did not vex me; it would not be heard through the wall. At length it ceased. The old man

was dead. I removed the bed and examined the corpse. Yes, he was stone, stone dead. I placed my hand upon the heart and held it there many minutes. There was no pulsation. He was stone dead. His eye would trouble me no more.

If still you think me mad, you will think so no longer when I describe the wise precautions I took for the concealment of the body. The night waned and I worked hastily, but in silence. First of all I dismembered the corpse. I cut off the head and the arms and the legs.

I then took up three planks from the flooring of the chamber, and deposited all . . . I then replaced the boards so cleverly, so cunningly, that no human eye—not even his—could have detected any thing wrong. There was nothing to wash out—no stain of any kind—no blood spot whatever. I had been too wary for that. A tub had caught all—ha! ha!

When I had made an end of these labours, it was four o'clock—still dark as midnight. As the bell sounded the hour, there came a knocking at the street door. I went down to open it with a light heart—for what had I now to fear? There entered three men, who introduced themselves with perfect suavity, as officers of the police. A shriek had been heard by a neighbour during the night; suspicion of foul play had been aroused; information had been lodged at the police office, and they had been deputed to search the premises.

I smiled—for what had I to fear? I bade the gentlemen welcome. The shriek, I said, was my own in a dream. The old man, I mentioned, was absent in the country. I took my visitors all over the house. I bade them search—search well. I led them, at length, to his chamber. I showed them his treasure, secure, undisturbed. In the enthusiasm of my confidence, I brought chairs into the room, and desired them here to rest from their fatigues, while I myself, in the wild audacity of my perfect triumph, placed my own seat upon the very spot beneath which reposed the corpse of the victim.

The officers were satisfied. My manner had convinced them. I was singularly at ease. They sat, and while I answered cheerily, they chatted familiar things. But, ere long, I felt myself getting pale and wished them gone. My head ached, and I fancied a ringing in my ears; but still they sat and still chatted. The ringing became more distinct: I talked more freely to get rid of the feeling: but it continued and gained definitiveness—until, at length, I found that the noise was not within my ears.

No doubt I now grew very pale—but I talked more fluently, and with a heightened voice. Yet the sound increased—and what could I do? It was a low, dull, quick sound—much such a sound as a watch makes when enveloped in cotton. I gasped for breath—and yet the officers heard it not. I talked more quickly—more vehemently; but the noise steadily increased. I arose and argued about trifles, in a high key and with violent gesticulations, but the noise steadily increased. Why would they not be gone? I paced the floor, to and fro, with heavy strides, as if excited to fury by the observations of the men—but the noise steadily increased. Oh God! What could I do? I foamed—I raved—I swore! I swung the chair upon which I had been sitting, and grated it upon the boards, but the noise arose over all and continually increased. It grew *louder—louder—louder—louder! And still the men chatted pleas-antly and smiled. Was it possible they heard not? Almighty God!—no, no! They heard! They suspected! They knew! They were making a mockery of my horror! This I thought and this I think. But anything was better than this agony! Anything was more tolerable than this derision! I could bear those hypocritical smiles no longer! I felt that I must scream or die!—and now—again—hark! louder! louder! louder! louder!*

"Villains!" I shrieked, "dissemble no more! I admit the deed! Tear up the planks! Here, here!—it is the beating of his hideous heart!"

IF I DID IT

Is the narrator here the protagonist or the antagonist? Or both? He is the major character (protagonist) and maybe his own conscience is the force against him (antagonist). Is he a man divided against himself? Maybe the old man's hideous eye is the antagonist and our guy is the protagonist.

Note how he, the murderer, takes great pains to tell us he's not a madman, because he is so methodical and rational. Ever hear Shakespeare's line about Lady Macbeth—*"She protesteth too much"*? Our guy here protests so much that he's sane, we are tempted to suspect he's crazy. What is his problem? Is that dead man's heart really beating? How can that be? The murderer felt it, and it was stone cold. What is going on?

YIKES! GOODS THIS SHORT STORY NEEDS

Something creepy. A stalking animal or person, a madman or woman, or even a child, an unnatural animal-person, an alien. This is the antagonist. And there must be a protagonist/victim, preferably someone sane, but frightened out of his boots. The scary thing that creates the horror can be a certifiably insane person, a temporarily mad person, or anything that threatens the protagonist's existence. The victim or victims can be dead at the start, in the middle, or at the end. This story can be told in first-person or third-person omniscient, or limited, or objective.

THE END IS AT HAND

Try this excerpt from another horror story, *Little Louise Roque*, by Guy de Maupassant, who wrote many wonderful short stories; this theme is similar to Poe's. Here, a 13-year-old girl is found by a postman horribly murdered in the mayor's woods. The postman runs to tell the mayor, M. Renardet. The mayor brings a doctor and other helpers. The girl is identified as Louise Roque, whose mother has a breakdown when she sees the body. M. Renardet sets out to find the murderer. He continues to be drawn back to the woods where the body was found. No one is able to solve the crime. Then:

Little Louise Roque

From the day when the investigation came to a close, he [M. Renardet] became gradually nervous, more excitable . . . Sudden noises made him jump up with fear; he shuddered at the slightest thing, trembled sometimes from head to foot when a fly alighted on his forehead. Then he was seized with an imperious desire for motion, which compelled him to keep continually on foot, and made him remain up whole nights walking to and fro in his own room . . . *Every moment his thoughts returned to that horrible scene, and, though he endeavored to drive away the picture from his mind, though he put it aside with terror, with disgust, he felt it surging through his soul, moving about in him, waiting incessantly for the moment to reappear.*

Then in the night, he was afraid, afraid of the shadows falling around him . . . he instinctively feared it, felt that it was peopled with terrors . . . the night, the impenetrable night, thicker than walls, and empty, the infinite night, so black, so vast, in which one might brush against frightful things, the night when one feels that mysterious terror is wandering, prowling about, appeared to him to conceal an unknown danger, close and menacing . . . Was it true that this curtain did move? He asked himself, fearing that his eyes had deceived him. It was, moreover, such a slight thing, a gentle flutter of lace . . . Renardet sat still, with staring eyes, and outstretched neck. Then he sprang to his feet abruptly ashamed of his fear, took four steps, seized the drapery with both hands, and pulled it wide apart . . . and suddenly perceived a light, a moving light, which seemed some distance away . . . Suddenly this light became an illumination, and he beheld little Louise Roque naked and bleeding on the moss. He recoiled frozen with horror, sank into his chair and fell backward . . . He had had a hallucination—that was all: a hallucination due to the fact that a marauder of the night was walking with a lantern . . . What was there astonishing, besides, in the circumstance that the recollection of his crime should somehow bring before him the vision of the dead girl? . . . He took off his clothes, extinguished the lamp, and lay down, shutting his eyes . . . By dint of straining his eyes, he could perceive some star, and he arose, groped his way across the room, discovered the panes with his outstretched hands, and placed his forehead close to them. There below, under the trees, the body of the little girl glittered like phosphorus, lighting up the surrounding darkness.

Renardet uttered a cry and rushed toward his bed, where he lay till morning, his head hidden under the pillow. From that moment, his life became intolerable . . .

Well, my goodness. Renardet seems mightily disturbed. What do you think might be wrong with him? Is it the same thing that's wrong with Poe's narrator in the earlier story? Note, too, toward the end, the narrator says *"the recollection of his crime,"* so the reader hears for the first time that he, M. Renardet, is the murderer.

As you can see, authors can manufacture horror stories in people's minds by using guilt. The tactics of Poe and de Maupassant put the reader directly into the heads of the murderers, so we feel their pain, as they increasingly realize the horrors they have wrought.

De Maupassant's story is in third person, limited omniscient. Does the pain of the man register in your mind as vividly and horrifically as Poe's story, told in first person? Notice what you, the writer, can do to tone by shifting the point of view.

WHAT IF

What if a scientist has a pet cat that eats something weird in his lab and turns into a monster cat that stalks a boy or girl or a whole town full of people? What if it gets so big that it's bigger than the houses? What if it can crush a house with its giant paw?

What if a man or woman kills someone unintentionally, but because these characters are somewhere they should not be (where and why?), the killer has to deny or repress the crime? What's in the killer's mind?

What if dating teenagers are curious about an old abandoned house at the end of a street and go inside to explore it? They run out screaming. What do they find that spooks them out?

What if a woman who has been away comes home to her house in the country on a stormy night? She's come home a day early to surprise her husband, but he's not there. The storm has knocked down trees and her phone line is out. She goes to the basement to get wood for a fire and finds a dead woman. No phone. No neighbors. No husband. Now what?

Scary.

CHAPTER 10

Twisted Skeletons:
Surprise-Ending Stories

THE **SURPRISE-ENDING** short story is one that's fun to play around with. Here, you can unzip your imagination and let it run rampant, because the story can be about most anyone and anything—families, mysteries, romance, adventure, love, hate, jealousy, horror, or a combination of these. These stories can have highly-placed people falling from their high places or lowly-placed people climbing from rags to riches. The only sure-fire ingredient in a recipe for this story is that in the end there must be a twist or surprise that catches the reader off-guard.

These stories are usually driven by Character, rather than Plot, because the reader needs to care about the protagonist to care about the story. And to care about the character, the reader needs to know a good deal about him or her, which means the writer has to develop that character carefully. The plot will be about how this character behaves, leading to the twist and surprise ending.

SHOCK IT TO 'EM: GOODS THIS SHORT STORY NEEDS

So, there has to be a main character and something against which he or she must fight or flee. This, of course, is necessary in all stories of any kind (and in dramatic plays, too). Sometimes that force is very strong—life-threatening to the main character. Sometimes, as in Hemingway's *Soldier's Home,* that force can be benign or subtle, but also deadly for the protagonist. If you remember, in Hemingway's story, the protagonist, Krebs, has already met with the fierce, life-threatening force—the war he fought in. Then he comes home and the force is just his sweet family who wants to love him and be loved by him. But the strongest antagonist, the war, has left him a changed man.

And in the end of a surprise-ending story either the protagonist or antagonist has to have changed in some way—one can twist into the other, they can change places, the protagonist can suddenly learn that he's fighting the wrong force, or, as in the story you will next read, the author plays with your head, so you don't know who is the good guy and who is the bad guy.

SATISFYING TWISTS

Many writers have written stories in this genre, but O. Henry was a master of the form. An interesting thing about the following story is that O. Henry himself (when his name was William Sidney Porter) went to federal prison for three years for bank fraud. A prison guard, Orrin Henry, had a big influence on Porter, who then decided to follow the straight and narrow. After he left prison, Porter took the name O. Henry, based on his mentor-guard's, and wrote *A Retrieved Reformation* and many other stories. His life, as well as his stories, had a surprise ending. This is the story, in its entirety.

A Retrieved Reformation

A guard came to the prison shoe-shop, where Jimmy Valentine was assiduously stitching uppers, and escorted him to the front office. There the warden handed Jimmy his pardon, which had been signed that morning by the governor. Jimmy took it in a tired kind of way. He had served nearly ten months of a four-year sentence. He had expected to stay only about three months, at the longest. When a man with as many friends on the outside as Jimmy Valentine had is received in the "stir" it is hardly worth while to cut his hair.

"Now, Valentine," said the warden, "you'll go out in the morning. Brace up, and make a man of yourself. You're not a bad fellow at heart. Stop cracking safes, and live straight."

"Me?" said Jimmy, in surprise. "Why, I never cracked a safe in my life."

"Oh, no," laughed the warden. "Of course not. Let's see, now. How was it you happened to get sent up on that Springfield job? Was it because you wouldn't prove an alibi for fear of compromising somebody in extremely high-toned society? Or was it simply a case of a mean old jury that had it in for you? It's always one or the other with you innocent victims."

"Me?" said Jimmy, still blankly virtuous. "Why, warden, I never was in Springfield in my life!"

"Take him back, Cronin," smiled the warden, "and fix him up with outgoing clothes. Unlock him at seven in the morning, and let him come to the bull-pen. Better think over my advice, Valentine."

At a quarter past seven on the next morning Jimmy stood in the warden's outer office. He had on a suit of the villainously fitting, ready-made clothes and a pair of stiff, squeaky shoes that the state furnishes to its discharged compulsory guests.

The clerk handed him a railroad ticket and the five-dollar bill with which the law expected him to rehabilitate himself into good citizenship and prosperity. The warden gave him a cigar, and shook hands. Valentine, 9762, was chronicled on the books "Pardoned by the Governor," and Mr. James Valentine walked out into the sunshine.

Disregarding the song of the birds, the waving green trees and the smell of the flowers, Jimmy headed straight for a restaurant. There he tasted the first sweet joys of liberty in the shape of a broiled chicken and a bottle of white wine—followed by a cigar a grade better than the one the warden had given him. From there he proceeded leisurely to the depot. He tossed a quarter into the hat of a blind man sitting by the door, and boarded his train. Three hours set him down in a little town near the state line. He went to the café of one Mike Dolan and shook hands with Mike, who was alone behind the bar.

"Sorry we couldn't make it sooner, Jimmy, me boy," said Mike. "But we had that protest from Springfield to buck against, and the governor nearly balked. Feeling all right?"

"Fine," said Jimmy. "Got my key?"

He got his key and went upstairs, unlocking the door of a room at the rear. Everything was just as he had left it. There on the floor was still Ben Price's collar-button that had been torn from that eminent detective's shirt-band when they had overpowered Jimmy to arrest him.

Pulling out from the wall a folding bed, Jimmy slid back a panel in the wall and dragged out a dust-covered suitcase. He opened this and gazed fondly at the finest set of burglar's tools in the East. It was a complete set, made of specially tempered steel, the latest designs in drills, punches, braces and bits, jimmies, clamps, and augers, with two or three novelties invented by Jimmy himself, in which he took pride. Over nine hundred dollars they had cost him to have made at___, a place where they make such things for the profession.

In half an hour Jimmy went downstairs and through the café. He was now dressed in tasteful and well-fitting clothes, and carried his dusted and cleaned suitcase in his hand.

"Got anything on?" asked Mike Dolan, genially.

"Me?" said Jimmy, in a puzzled tone. "I don't understand. I'm representing the New York Amalgamated Short Snap Biscuit Cracker and Frazzled Wheat Company."

This statement delighted Mike to such an extent that Jimmy had to take a seltzer and milk on the spot. He never touched "hard" drinks.

A week after the release of Valentine, 9762, there was a neat job of safe-burglary done in Richmond, Indiana, with no clue to the author. A scant eight hundred dollars was all that was secured. Two weeks after that a patented, improved burglar-proof safe in Logansport was opened like a cheese to the tune of fifteen hundred dollars, currency; securities and silver untouched. That began to interest the rogue-catchers. Then an old-fashioned bank safe in Jefferson City became active and threw out of its crater an eruption of bank-notes amounting to five thousand dollars. The losses were now high enough to bring the matter up into Ben Price's class of work. By comparing notes, a remarkable similarity in the methods of the burglaries was noticed. Ben Price investigated the scenes of the robberies, and was heard to remark:

"That's Dandy Jim Valentine's autograph. He's resumed business. Look at that combination knob—jerked out as easy as pulling up a radish in wet weather. He's got the only clamps that can do it. And look how clean those tumblers were punched out! Jimmy never has to drill but one hole. Yes, I guess I want Mr. Valentine. He'll do his bit next time without any short-time or clemency foolishness."

Ben Price knew Jimmy's habits. He had learned them while working up the Springfield case. Long jumps, quick get-aways, no confederates, and a taste for good society—these ways had helped Mr. Valentine to become noted as a successful dodger of retribution. It was given out that Ben Price had taken up the trail of the elusive cracksman, and other people with burglar-proof safes felt more at ease.

One afternoon Jimmy Valentine and his suitcase climbed out of the mail-hack in Elmore, a little town five miles off the railroad down *in the black-jack country of Arkansas. Jimmy, looking like an athletic young senior just home from college, went down the board sidewalk toward the hotel.*

A young lady crossed the street, passed him at the corner and entered the door over which was the sign, "The Elmore Bank." Jimmy Valentine looked into her eyes, forgot what he was, and became another man. She lowered her eyes and colored slightly. Young men of Jimmy's style and looks were scarce in Elmore.

Jimmy collared a boy that was loafing on the steps of the bank as if he were one of the stockholders, and began to ask him questions about the town, feeding him dimes at intervals. By and by the young lady came out, looking royally unconscious of the young man with the suit-case, and went her way.

"Isn't that young lady Miss Polly Simpson?" asked Jimmy, with specious guile.

"Naw," said the boy. "She's Annabel Adams. Her pa owns this bank. What'd you come to Elmore for? Is that a gold watch chain? I'm going to get a bulldog. Got any more dimes?"

Jimmy went to the Planters' Hotel, registered as Ralph D. Spencer, and engaged a room. He leaned on the desk and declared his platform to the clerk. He said he had come to Elmore to look for a location to go into business. How was the shoe business now, in the town? He had thought of the shoe business. Was there an opening?

The clerk was impressed by the clothes and manner of Jimmy. He, himself, was something of a pattern of fashion to the thinly gilded youth of Elmore, but he now perceived his shortcomings. While trying to figure out Jimmy's manner of tying his four-in-hand, he cordially gave information.

Yes, there ought to be a good opening in the shoe line. There wasn't an exclusive shoe-store in the place. The dry-goods and general stores handled them. Business in all lines was fairly good. Hoped Mr. Spencer would decide to locate in Elmore. He would find it a pleasant town to live in, and the people very sociable.

Mr. Spencer thought he would stop over in the town a few days and look over the situation. No, the clerk needn't call the boy. He would carry up his suit-case, himself; it was rather heavy.

Mr. Ralph Spencer, the phoenix that arose from Jimmy Valentine's ashes—ashes left by the flame of a sudden and alterative attack of love—remained in Elmore, and prospered. He opened a shoe-store and secured a good run of trade.

Socially, he was also a success, and made many friends. And he accomplished the wish of his heart. He met Miss Annabel Adams, and became more and more captivated by her charms.

At the end of a year, the situation of Mr. Ralph Spencer was this: he had won the respect of the community, his shoe-store was flourishing, and he and Annabel were engaged to be married in two weeks. Mr. Adams, the typical, plodding, country banker, approved of Spencer. Annabel's pride in him almost equalled her affection. He was as much at home in the family of Mr. Adams and that of Annabel's married sister as if he were already a member.

One day Jimmy sat down in his room and wrote this letter, which he mailed to the safe address of one of his old friends in St. Louis:

> *Dear Old Pal:*
>
> *I want you to be at Sullivan's place, in Little Rock, next Wednesday night at nine o'clock. I want you to wind up some little matters for me. And, also, I want to make you a present of my kit of tools. I know you'll be glad to get them—you couldn't duplicate the lot for a thousand dollars. Say, Billy, I've quit the old business—a year ago. I've got a nice store. I'm making an honest living, and I'm going to marry the finest girl on earth two weeks from now. It's the only life, Billy—the straight one. I wouldn't touch a dollar of another man's money now for a million. After I get married I'm going to sell out and go West, where there won't be so much danger of having old scores brought up against me. I tell you, Billy, she's an angel. She believes in me; and I wouldn't do another crooked thing for the whole world. Be sure to be at Sully's, for I must see you. I'll bring along the tools with me.*
>
> *Your old friend,*
> *Jimmy*

On the Monday night after Jimmy wrote this letter, Ben Price jogged unobtrusively into Elmore in a livery buggy. He lounged about town in his quiet way until he found out what he wanted to know. From the drug store across the street from Spencer's shoe-store he got a good look at Ralph D. Spencer.

"Going to marry the banker's daughter, are you, Jimmy?" said Ben to himself, softly. "Well, I don't know!"

The next morning Jimmy took breakfast at the Adamses. He was going to Little Rock that day to order his wedding-suit and buy something nice for Annabel. That would be the first time he had left town since he came to Elmore. It had been more than a year now since those last professional "jobs," and he thought he could safely venture out.

After breakfast quite a family party went downtown together—Mr. Adams, Annabel, Jimmy, and Annabel's married sister with her two little girls, aged five and nine. They came by the hotel where Jimmy still boarded, and he ran up to his room and brought along his suit-case. Then they went to the bank. There stood Jimmy's horse and buggy and Dolph Gibson, who was going to drive him over to the railroad station.

All went inside the high, carved oak railings into the banking-room— Jimmy included, for Mr. Adams's future son-in-law was welcome anywhere. The clerks were pleased to be greeted by the good-looking, agreeable young man who was going to marry Miss Annabel. Jimmy set his suit-case down. Annabel, whose heart was bubbling along with happiness and lively youth, put on Jimmy's hat and picked up the suit-case. "Wouldn't I make a nice drummer?" said Annabel. "My! Ralph, how heavy it is. Feels like it's full of gold bricks."

"Lot of nickel-plated shoe-horns in there," said Jimmy, coolly, "that I'm going to return. Thought I'd save express charges by taking them up. I'm getting awfully economical."

The Elmore Bank had just put in a new safe and vault. Mr. Adams was very proud of it, and insisted on an inspection by every one. The vault was a small one, but it had a new patented door. It fastened with three solid steel bolts, thrown simultaneously with a single handle, and had a time-lock. Mr. Adams beamingly explained its workings to Mr. Spencer, who showed a courteous but not too intelligent interest. The two children, May and Agatha, were delighted by the shining metal and funny clock and knobs.

While they were thus engaged Ben Price sauntered in and leaned on his elbow, looking casually inside between the railings. He told the teller that he didn't want anything, he was just waiting for a man he knew.

Suddenly there was a scream or two from the women, and a commotion. Unperceived by the elders, May, the nine-year-old girl, in a spirit of play, had shut Agatha in the vault. She had then shot the bolts and turned the knob of the combination as she had seen Mr. Adams do.

The old banker sprang to the handle and tugged at it for a moment. "The door can't be opened," he groaned. "The clock hasn't been wound nor the combination set."

Agatha's mother screamed again, hysterically.

"Hush!" said Mr. Adams, raising his trembling hand. "All be quiet for a moment. Agatha!" he called as loudly as he could. "Listen to me." During the following silence they could just hear the faint sound of the child wildly shrieking in the dark vault in a panic of terror.

"My precious darling!" wailed the mother. "She will die of fright! Open the door! Oh, break it open! Can't you men do something?"

"There isn't a man nearer than Little Rock who can open that door," said Mr. Adams, in a shaky voice. "My God! Spencer, what shall we do? That child—she can't stand it long in there. There isn't enough air, and, besides, she'll go into convulsions from fright."

Agatha's mother, frantic now, beat on the door of the vault with her hands. Somebody wildly suggested dynamite. Annabel turned to Jimmy, her large eyes full of anguish, but not yet despairing. To a woman nothing seems quite impossible to the powers of the man she worships.

"Can't you do something, Ralph—try, won't you?"

He looked at her with a queer, soft smile on his lips and in his keen eyes.

"Annabel," he said, "give me that rose you are wearing, will you?"

Hardly believing that she heard him aright, she unpinned the bud from the bosom of her dress, and placed it in his hand. Jimmy stuffed it into his vest-pocket, threw off his coat and pulled up his shirt-sleeves. With that act Ralph D. Spencer passed away and Jimmy Valentine took his place.

"Get away from the door, all of you," he commanded, shortly.

He set his suit-case on the table and opened it out flat. From that time on he seemed to be unconscious of the presence of any one else. He laid out the shining, queer implements swiftly and orderly, whistling softly to himself as he always did when at work. In a deep silence and immovable, the others watched him as if under a spell.

In a minute Jimmy's pet drill was biting smoothly into the steel door. In ten minutes—breaking his own burglarious record—he threw back the bolts and opened the door.

Agatha, almost collapsed, but safe, was gathered into her mother's arms.

Jimmy Valentine put on his coat, and walked outside the railings towards the front door. As he went he thought he heard a far-away voice that he once knew call "Ralph!" But he never hesitated.

At the door, a big man stood somewhat in his way.

"Hello Ben!" said Jimmy, still with his strange smile. "Got around at last, have you? Well, let's go. I don't know that it makes much difference, now."

And then Ben Price acted rather strangely.

"Guess you're mistaken, Mr. Spencer," he said. "Don't believe I recognize you. Your buggy's waiting for you, ain't it?"

And Ben Price turned and strolled down the street.

PLAYING IT SAFE OR PLAYING A SAFE?

Well, were you surprised? Maybe not surprised that Ralph/Jimmy got out his tools to save the girl (because the author has given him some honorable qualities—he wants to go straight, he's likable, and he's sweet to his loved one), even though he knows his tools and vault-cracking will tip his hand and lose him his fiancé. But you were probably surprised that Ben Price let him go.

Jimmy is set up as the "bad guy" here—a law-breaker, a jail bird who games the system with "high-toned" friends to get him a pardon. And Ben Price is set up as the "good guy" out to get him—the force against him, or the antagonist.

Jimmy does go straight. Falls in love and is in line to become one of those high-toned men himself. Then he is tested, as love is always tested. He passes the test as he puts his love on the line. The little girl is saved, and Jimmy becomes the "good guy," while the detective stalking him in the lobby becomes the "bad guy."

Then Ben lets Jimmy pass right by and the good and bad guys change hats again. Ben is "good" for realizing that Jimmy is a decent guy, trying to go straight, and who blows it all to save a little girl. Jimmy is more of a mystery. Is he going off with his tools in his hand to resume his old "bad guy" criminal life? Will Annabel rush out to stop him? Will he confess his past to her? If so, will she forgive him and profess to love him anyway? Will banker father approve of that? Or will Jimmy simply walk off into the sunset . . . and we don't know what?

The ending is a surprise, and surprisingly ambiguous.

By the way, why did Jimmy ask for the rose from his love? Remember that short stories are so short every word counts. Nothing in them is casual. There must be a reason Jimmy did that . . .

(THIS JUST IN: O. Henry never got in trouble with the law again, as far as we know . . . which should show you not to read too much biography of an author into a story.)

Another story with a twist is called *The Landlady* by Roald Dahl (whom you may know as having written a lot of eerie stories for children). In this story, a young man, Billy Weaver, gets out of a train in an unfamiliar small town in England, where he hopes to get a job. He finds an enticing-looking bed and breakfast, looks through the window and sees a fireplace, a sleeping dog before it, a colorful parrot in a cage. It all looks warm and cozy, so he goes in to get a room for the night. The landlady seems slightly nutty, but harmless and sweet. She offers him tea, and he signs the guest book. In it he sees two names of young men that look familiar to him. The landlady chats with him, while he notices the dog doesn't move and the parrot doesn't squawk. The landlady tells him in her innocent voice that the two men (whose names Billy has now remembered in reading of their mysterious disappearance in the newspaper) are still residing with her upstairs. The more Billy sits and observes and listens, the more he realizes that the landlady is a taxidermist, who poisons her guests, along with her dog and parrot, then stuffs them and keeps them.

But Billy has already drunk his tea . . .

The story becomes more and more creepy as you read, and then you, the reader, are hit with the truth of the situation just about the time Billy is. This short story with a twist is a little mystery we call a *cozy*—because the whole scene is tightly drawn and indeed cozy, has a dotty little woman in it, and the action creeps up on the reader.

WHAT IF

What if a wife and mother, Susan, tells her husband, Mark, who works at home, that she's joined a gym and will go off for three days a week for two hours so she can remain strong and healthy for her family? She goes. After two months, Mark gets a call that Susan's sister is very sick in a place far away and Susan needs to go to her. Mark goes to the gym to tell Susan. She's not there. She's not even registered. Mark goes off frantically convincing himself that she's having a love affair. With whom? Why? Meanwhile, where is Susan? What's the backstory?

What if Paul, a golf caddy, steals Mary away from a haughty rich boy at the country club? They run away together to go sailing and get swept away by a squall; their boat is destroyed. They finally reach a small island. How do they get there? How long do they have to stay before they're rescued? Who rescues them? What happens after that?

What if high school seniors, Jennifer and Brad, are inseparable? Then they both get worried that the other will go off to college and forget him or her. Brad becomes more and more unavailable to go out with Jennifer. Jennifer starts flirting with other boys. Brad hears about it. Why is Brad unavailable? Why is Jennifer flirting?

What if two bumbling "bad guys" kidnap a little boy to get ransom money from his rich dad? The little boy is impossible—he acts out, is hyperactive, and drives the kidnappers nuts. Where do they take him? How do they treat him? What do they finally do with him? (O. Henry has written this story already, but write your own. Then look up *The Ransom of Red Chief.*)

CHAPTER 11

Double Take. Whammy.

OKAY, YOU CHEATER! Has the DOUBLE TAKE, WHAMMY hit yet? Good, you can stop reading right here. (Who needs to be in Chapter 11, anyway?) No? Try it again. Think of "time" as the verb and "flies" as the noun. Now, got it? Ohhhhh. . . . OK. You're going to write a good short story yet—maybe one with a surprise ending.

NOTES UNZIPPED

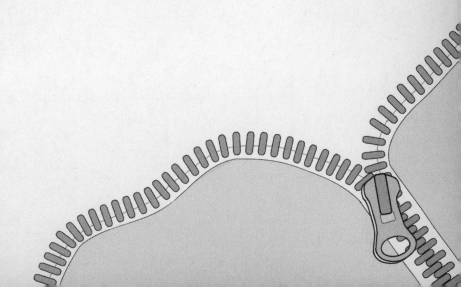

NOTES UNZIPPED

NOTES UNZIPPED

NOTES UNZIPPED

GIVE US YOUR FEEDBACK

Peterson's, a Nelnet company, publishes a full line of resources to help guide you through the college admission process. Peterson's publications can be found at your local bookstore, library, and high school guidance office, and you can access us online at www.petersons.com.

We welcome any comments or suggestions you may have about this publication and invite you to complete our online survey at www.petersons.com/booksurvey. Or you can fill out the paper survey on the next page, tear it out, and mail it to us at:

Publishing Department
Peterson's
2000 Lenox Drive
Lawrenceville, NJ 08648

Peterson's
Book Satisfaction Survey

ive Us Your Feedback

ınk you for choosing Peterson's as your source for personalized solutions for your education and career ıievement. Please take a few minutes to answer the following questions. Your answers will go a long way in ping us to produce the most user-friendly and comprehensive resources to meet your individual needs.

ıen completed, please tear out this page and mail it to us at:

Publishing Department
Peterson's, a Nelnet company
2000 Lenox Drive
Lawrenceville, NJ 08648

ıu can also complete this survey online at **www.petersons.com/booksurvey.**

What is the ISBN of the book you have purchased? (The ISBN can be found on the book's back cover in the lower right-hand corner.) _____

Where did you purchase this book?
❑ Retailer, such as Barnes & Noble
❑ Online reseller, such as Amazon.com
❑ Petersons.com
❑ Other (please specify) _____

If you purchased this book on Petersons.com, please rate the following aspects of your online purchasing experience on a scale of 4 to 1 (4 = Excellent and 1 = Poor).

	4	3	2	1
Comprehensiveness of Peterson's Online Bookstore page	❑	❑	❑	❑
Overall online customer experience	❑	❑	❑	❑

Which category best describes you?
❑ High school student
❑ Parent of high school student
❑ College student
❑ Graduate/professional student
❑ Returning adult student

❑ Teacher
❑ Counselor
❑ Working professional/military
❑ Other (please specify) _____

Rate your overall satisfaction with this book.

Extremely Satisfied	Satisfied	Not Satisfied
❑	❑	❑

6. Rate each of the following aspects of this book on a scale of 4 to 1 (4 = Excellent and 1 = Poor).

	4	3	2	1
Comprehensiveness of the information	❏	❏	❏	❏
Accuracy of the information	❏	❏	❏	❏
Usability	❏	❏	❏	❏
Cover design	❏	❏	❏	❏
Book layout	❏	❏	❏	❏
Special features (e.g., CD, flashcards, charts, etc.)	❏	❏	❏	❏
Value for the money	❏	❏	❏	❏

7. This book was recommended by:
- ❏ Guidance counselor
- ❏ Parent/guardian
- ❏ Family member/relative
- ❏ Friend
- ❏ Teacher
- ❏ Not recommended by anyone—I found the book on my own
- ❏ Other (please specify) _____

8. Would you recommend this book to others?

Yes	Not Sure	No
❏	❏	❏

9. Please provide any additional comments.

_____

_____

_____

_____

_____

Remember, you can tear out this page and mail it to us at:

> Publishing Department
> Peterson's, a Nelnet company
> 2000 Lenox Drive
> Lawrenceville, NJ 08648

or you can complete the survey online at **www.petersons.com/booksurvey.**

Your feedback is important to us at Peterson's, and we thank you for your time!

If you would like us to keep in touch with you about new products and services, please include your e-mail address here: _____